With Good Reason

AN INTRODUCTION TO INFORMAL FALLACIES

With Good Reason

AN INTRODUCTION TO INFORMAL FALLACIES
Fourth Edition

S. Morris Engel
York University

ST. MARTIN'S PRESS   NEW YORK

Editor: Debra Nesbitt
Project editor: Robert Skiena
Production associate: Katherine Battiste
Cover design: Darby Downey
Cover art: Eldon Doty

Library of Congress Catalog Card Number: 88-63086

For information, write:
St. Martin's Press, Inc.
175 Fifth Avenue
New York, NY 10010

ISBN: 0-312-02130-5

The drawing on the cover illustrates the Fallacy of Amphiboly.

**For
My Wife**

Preface

The text that readers now have before them is in tone and structure essentially the same as its immediate predecessor. As in the previous revision, most of the changes consist of further textual revisions or additions.

The most extensive of these consist of the inclusion of the fallacy of slippery slope, the revision of several exercises, and the addition, in an appendix, of a new section on writing, particularly the writing of an argumentative essay. This last is a major change and a major new emphasis. I hope it will be found to be an effective aid in helping students improve their powers of thought and expression. Depending upon the class's previous preparation, the appendix can be taken up either simultaneously with part one of the text (with the early sections being assigned for independent study or review), or, if the class's preparation is weaker, upon the completion of part one, after the class has mastered the rudiments of logic and language and completed, in addition, chapter 3 on the fallacies of ambiguity.

The first edition of *With Good Reason* grew out of my concern that the students taking my introductory course in logic have a basis for evaluating the soundness of arguments which confront them at every turn and which they themselves continually devise. When I gave substantially the same course on television as part of the university's continuing education series, I was surprised by the enthusiastic response of the viewing audience. The members of the community seemed just as interested as my students on the campus in having information that would enable them to decipher the arguments of advertisers, politicians, their children, their business associates, and their friends. I began to suspect that the medieval trivium of logic, grammar, and rhetoric still had much to offer us.

The core of the text is the informal fallacies that are the subject of part two. Part one is preparatory, presenting the fundamentals of argument in nontechnical terms, plus those aspects of language, such as ambiguity, that can contribute to muddled thinking and muddled communication. Examples in the text are drawn from the mass media, from literature, from philosophic sources, and occasionally from my own invention. The story that concludes the text part two provides an amusing fictional treatment of the concepts the reader has encountered in the text. I have attempted throughout the text and in the exercises in each chapter to strike a balance between classical illustrations and those that have their basis in contemporary issues and objectives. My hope is that, through exposure to a variety of examples, readers will be better equipped to detect faulty reasoning in oral and written messages they encounter outside the text.

Unlike traditional treatments that reduce an argument to its essential structure in order to determine whether it is logically valid, the position underlying this text is that such reductions may fail to do justice to the full force or design of an argument. My experience with students has been that, although the validity of an argument depends on its form, what may initially appeal in an argument, and thereby prevent the detection of its weakness, are certain informal features. Thus an argument may appear valid because its formal invalidity is obscured as a result of vague or ambiguous language (giving rise to what are called fallacies of ambiguity in chapter 3); because it apes valid forms of argument (resulting in fallacies of presumption, which are discussed in chapter 4); or because the language used is emotionally appealing (producing fallacies of relevance, which are covered in chapter 5). The appeal of an argument, then, depends not exclusively on its structure as such but also on the use to which that structure is put. Frequently, as we shall see, the author of an argument uses a particular structure merely as a peg or diversion, while pressing his or her point home by a rather different route.

This work has grown slowly over a number of years and through many alterations. It owes an enormous debt to hundreds of authors whose books I read during the course of those years and to count-less students whose enthusiasm, suggestions, and criticisms contributed to every page. I wish to acknowledge again the help of my dear friend and former student Dr. Robert Archer, who has used this text in his own classes ever since it first appeared and has spared no pains in helping me make it as fine a work as possible. I am also enormously grateful to such further dear, former students and assistants as Doug Deaver, Frans Baert, and Sam Russo, and such friends and colleagues as Professors Henryk Skolimow-ski and K. T. Fann. I am enormously grateful to them all.

I would also like to thank here once again for their help and advice not only all the people who helped me with the three previous editions of this text but all those who have now joined them in this new revision. For taking such great care with this work and going so deeply into it, I would like to thank Professors Larry Hitterdale, Charlie Busch, Jr., Edward L. Schoen, Leland Creer, Robert Gurland, John King, John Meixner, Nancy Nersessian, Mary Philips, David Weinberger, and Michael Myers. And for help with this new revision, many thanks to the following reviewers for St. Martin's Press: William Brown, Southwest Missouri State University; Gary L. Cesarz, Ph.D., University of New Mexico; William Drumin, Ph.D., King's College; Lance R. Factor, Knox College; Patricia Fauser, Illinois Benedictine College; Sidney Gendin, Ph.D., Eastern Michigan University; David E. Gibson, Ph.D., Pepperdine University; John Meixner, Central Michigan University; Ronald K. Messer, Ricks College; Stephen Pollard, Northeast Missouri State University; Walter A. Shelburne, Ph.D., San Jose State University, and David L. Treybig, Ph.D., Baldwin-Wallace College. As in the past, I have tried to be guided by them, and I hope they will not find the results disappointing.

As in the previous three editions, solutions to a number of exercises are given in the text. The instructor's manual, which is bound into the back of the instructor's edition, contains solutions to those examples not solved in the text, as well as additional examples for class use.

I am grateful again to all for the warm reception accorded *With Good Reason* and the success it has had. I hope it will continue to be deserving of it.

S. MORRIS ENGEL

Contents

Part One

On Logic and Language

Part one of our study of informal logic presents some basic principles and considerations that will be necessary for analysis of the common logical errors treated in part two. This book is thus organized on the assumption that logic is similar to a science such as medicine in that, for its full understanding, we require a knowledge of correct functioning before we can fully appreciate what often goes wrong. Accordingly, the first chapter sets forth the fundamentals of logical argument: what a good argument does and the attributes it must possess. Here we shall explore the nature of reasoning.

But reason is only one of two tools used in argument, the other being language. As we reason through a problem, we put our reasoning into words—in speaking, in writing, or silently to ourselves. Because language plays such a vital role in argument, chapter 2 undertakes to sensitize the reader to the capacity that words have to enrich, distort, specify, obscure, sharpen, or confuse our thinking. We shall see that precise use of language is closely tied to precise use of reason.

With correct reasoning and appropriate language established as cornerstones, we shall be prepared to go on, in part two, to analyze typical mistakes in logic. The reader will doubtless not be far along in chapter 1 before realizing that formulation of a good argument is not always an easy matter—that, indeed, much of what we hear, read, say, and think every day is less than correct from a logical point of view. This has always been true, as the ancient Greeks recognized when they began the study of logic as we know it.

To assist in their understanding of logical reasoning and to come to grips with common errors in logic, the Greeks began to catalog various logical errors. Those who followed them continued the study—often in Latin, which provides many of the words and phrases still used by logicians. We shall pause at various points in this text to note briefly the history of important words and phrases which have come down to us from Greek and Latin. We shall see that these unfamiliar phrases name errors that are just as prevalent today as they were in ancient Greece or Rome. We have as much need to study them as did the early philosophers, and this study will form the subject of part two.

We have likened the study of logic to medical science, in that just as we need to understand the workings of a healthy, normal organism before we have a basis for studying the deviations from normality that are disease, so we need to understand correct logic before we can detect logical errors. We might even extend this comparison further, noting that just as the study of disease usually proves fascinating to beginning students of medicine, so most students taking a first course in informal logic find the subject as stimulating as any game.

Although studying logic is often enjoyable, we need to remember, however, that errors in logic can cause serious problems—for individuals, for groups, and for nations. The fruits of human reasoning on profound issues are matters of no little consequence, and several of the arguments examined in this text will assuredly strike the reader with their potential for harm. Some others, illustrating their points in an amusing way, may seem frivolous. Most of the arguments are taken from everyday situations, although some are invented and others have come down to us as classic illustrations in the history of logic. All reward close analysis by sharpening our ability to recognize bad arguments and increasing our appreciation of good ones.

"'You have a small capacity for reason, some basic tool-making skills, and the use of a few simple words.' . . . Yep. That's you."

1

The Nature and Scope of Logic

Anyone reading this book is already familiar with the seemingly constant battle for our minds and allegiances that has been such a distinctive feature of life in the last quarter of the twentieth century. Through the mass media particularly, we are bombarded with appeals to buy this product or that one, to believe this speaker or that one, to take this action or that. Persuasive messages come at us from our friends, from government, from all manner of sales strategies, from persons heatedly espousing public causes, even from persons with whom we engage in only casual discussion.

How are we to know whether to buy, believe, or do what is urged on us? What reasons exist, and how compelling are they? To what extent, if any, do they obligate us? And how can we better ensure that our own private analysis of problems that concern us is as reasonable as we can make it? One purpose of a study of logic is to gain tools with which to distinguish good arguments from bad ones. Accordingly, logic may be regarded as among the most powerful studies one can undertake, particularly in an age like ours which is so full of impressive claims and counterclaims.

1. LOGIC AS SCIENCE AND ART

Is logic a science (like astronomy or genetics), or is it a practical art (like gymnastics or cooking)? Is its object to describe the nature and structure of correct thinking, in the manner of an exact science? Or does it teach us how to reason correctly, as we might instruct someone in playing the trombone? Is its primary object, in short, to help us understand what clear reasoning is or to teach us to reason clearly?

A case can be made for viewing logic in either of these ways. Some have argued that logic is nothing but a science, in that it investigates, systematizes, and demonstrates rules of correct reasoning. They have even suggested that the teaching of logical reasoning may be pointless, just as we need not wait for the physiologist to teach us how to eat. The great British empiricist John Locke expressed this view when he remarked, "God has not been so sparing to men as to make them barely two-legged creatures, and left it to Aristotle to make them rational" (*An Essay Concerning Human Understanding,* bk. 4, ch. 17). Either we already know how to reason or we do not. If we already possess this faculty, we need no instruction. If we do not possess it, instruction will not help.

Others have argued that logic's main value is practical, in that it improves our reasoning powers and strengthens our ability to evaluate the correctness of arguments and to detect their weaknesses. Having such utility, they say, logic must be considered an art as well as a science for it not only informs the mind but trains it as well. Some have termed logic a *liberal* art, in that its study provides a better understanding of our nature and helps free us from ignorant thoughts and actions. ("Liberal" probably derives from the Latin *liber,* meaning "free.") In this book we shall follow the practice of those who regard the practical study of logic as of equal importance to its theoretical study. Our discussion will stress the recognition, analysis, and evaluation of arguments having application to the world of everyday affairs.

2. LOGIC AS THE STUDY OF ARGUMENT

Logic is the study of *argument.* As used in this sense, the word means not a quarrel (as when we "get into an argument") but a piece of reasoning in which one or more statements are offered as support for some other statement. The statement being supported is the *conclusion* of the argument. The reasons given in support of the conclusion are called *premises.* We may say, "This is so (conclusion) because that is so (premise)." Or, "This is so and this is so (premises); therefore that is so (conclusion)." Premises are generally preceded by such words as *because, for, since, on the ground that, in as much as,* and the like. Conclusions, on the other hand, are generally preceded by such words as *therefore, hence, consequently, it follows that, thus, so, we may infer that,* and *we may conclude that.*

The first step toward understanding arguments, therefore, is learning to identify premises and conclusions. To do so, look for the indicator words, as they are called, just listed. In arguments where such indicator words are absent, try to find the conclusion by determining what is the main thrust of the argument: the point the argument is trying to establish. That will be its conclusion; the rest its supporting grounds or premises.

Distinguishing the conclusion from the premise or premises in the following two arguments is easy, for in the first case one of its statements is preceded by the word *for* (which tells us that what follows is a premise and what remains must be its conclusion) and in the second, one of its statements is preceded by the word *hence* (which tells us that what follows is a conclusion and what remains must be its premise):

> a) Jones will not do well in this course, for he is having a hard time concentrating on schoolwork this semester and has hardly attended any classes.
> b) She has antagonized nearly everyone on the committee; hence it is unlikely that she will be granted the promotion.

In the following two examples, however, no such helpful indicator words are present:

> c) There are no foxes in this area. We haven't seen one all day.
> d) All communists favor public housing; Senator Smith favors it; he must be a communist.

To distinguish the premise from the conclusion in cases of this sort, ask yourself such questions as: what is being *argued for* and what is the person trying to *persuade us of?* In case *c* what is being argued for is not that "we haven't seen a fox all day"— for the other person obviously already knows this and is simply being reminded of it—but rather that, in light of this known fact, there *must* be no foxes in this area. *That* is the conclusion of the argument. Similarly with example *d:* what is being argued for is not that "all communists favor public housing," nor that "Senator Smith favors it"—for in this argument these are assumed to be shared statements of fact and stated as such—but rather that, in the light of these facts, Smith must be a communist.

The following is a somewhat more difficult example:

> e) It is an incontestable fact that our country is the ultimate, priceless goal of international communism. The leaders of international communism have vowed to achieve world domination. This cannot be until the Red Flag is flown over the United States. (J. Edgar Hoover on communism)

To distinguish the premises from the conclusion in this argument, ask yourself, what is the main thrust of this argument? Is it to persuade us of the fact that "the leaders of international communism have vowed to achieve world domination"? This hardly seems so, being offered, as it is, as a simple statement of fact. Is it that such world domination cannot be achieved "until the Red Flag is flown over the United States"? Again, hardly so, since this, too, is offered as a presumed fact, one entailed by the statement which immediately precedes it. We come then to the remaining statement, the one with which the argument opens: that it is "an incontestable fact that our country is the ultimate, priceless goal of international communism." Despite the use of the rhetorical phrase *incontestable fact,* what we have here is not a statement of fact but a conclusion gathered from the facts offered. Both from its tone and content this is what the argument aimed to establish and convince us of.

Finding the conclusion where it is not already indicated as such will not always be easy or certain. Our best aid will be attending carefully to the content and tone of the argument and to the direction of its reasoning.

EXERCISES

Distinguish between the conclusion and the premises in the following arguments.

* 1.   Since all rational beings are responsible for their actions, and since all human beings are rational, it follows that all human beings are responsible for their actions.

   2.   Jones does not attend church, for he is an atheist and atheists do not attend church.

   3.   It took over a year to finish the post office building; the plaster began to crack after it was completed; the heating system broke down. Someone must have been taking graft .

   4.   Dropping the bomb on Hiroshima caused great suffering. We ought not to have done it.

* 5.   The city should reimburse him for his hospital expenses for the simple reason that the accident took place while he was engaged on city business.

   6.   Because only those who can quote large chunks of that material can pass a test on it, it is useless for me to try, for I know hardly any of it by heart.

   7.   Today I will be master of my emotions. The tides advance; the tides recede. Winter goes and summer comes. Summer wanes and the cold increases. The sun rises; the sun sets. The moon is full; the moon is black. The birds arrive; the birds depart. Flowers bloom; flowers fade. Seeds are sown; harvests are reaped. All nature is a circle of moods and I am a part of nature and so, like the tides, my moods will rise; my moods will fall. Today I will be master of my emotions. (Og Mandino, *The Greatest Salesman in the World*)

   8.   Smith ought to exercise more. It would be good for his condition.

* 9.   Tom must have left already. He doesn't answer his phone.

   10. I no longer believe those who say that a poor politician could be a good President, "if he could only be appointed to the job." Without the qualities required of a successful candidate—without the ability to rally support, to understand the public, to express its aspirations—without the

organizational talent, the personal charm, and the physical stamina required to survive the primaries, the convention, and the election—no man would make a great President, however wise in other ways he might be. (Theodore G. Sorensen, *Decision-Making in the White House*)

11. The day *may* come when the rest of the animal creation may acquire those rights which never could have been withholden from them but by the hand of tyranny. The French have already discovered that the blackness of the skin is no reason why a human being should be abandoned without redress to the caprice of a tormentor. It may one day come to be recognized that the number of the legs, the villosity of the skin, or the termination of the *os sacrum,* are reasons equally insufficient for abandoning a sensitive being to the same fate. What else is it that should trace the insuperable line? Is it the faculty of reason, or perhaps the faculty of discourse? But a full-grown horse or dog is beyond comparison a more rational, as well as a more conversable animal, than an infant of a day, or a week, or even a month, old. But suppose they were otherwise, what would it avail? The question is not, Can they *reason?* nor Can they *talk?* but, Can they *suffer?* (Jeremy Bentham, *The Principles of Morals and Legislation,* 1789)

3. ARGUMENTS AND NONARGUMENTS

As we have seen, an argument is a piece of reasoning in which one or more statements are offered as support for some other statement. If a piece of writing makes no claim supported by such reasons it is not an argument. Thus questions are not arguments, nor are announcements, complaints, compliments, or apologies.

Such passages are not arguments because, again, they make no effort to persuade by offering reasons in support of their claims. For example, "Are there any plans to put 'The Little Rascals' on the air again?" is merely to pose a question, not an argument. It requests information, not assent to some claim.

The same is true of the following representative types of communications we all engage in from time to time:

a) I'm not going to watch anymore TV programs with laugh tracks. There's laughter if someone shuts a door. I'll laugh when I want to laugh. I think they should put the laugh tracks on the evening news when the weather forecasters are on the air.

b) I spent $125 to attend a reincarnation seminar and the leader appeared in a racing jacket, jeans, and a T-shirt advertising a California guitar shop. I consider that bad taste in Philadelphia. He is certainly the best regressionist I've seen in my sixty years, but you can have his Kung-Fu approach to spirituality.

c) The sincerest satisfaction in life comes in doing and not in dodging duty; in meeting and solving problems, in facing facts, in being a dependable person.

Example *a* is an expression of contempt and disgust, *b* is a complaint, and *c* merely a statement of a point of view without any attempt either to argue or persuade us of it.

None of these, then, are arguments. This does not mean that such passages are bad arguments; it simply means that they are not arguments at all. They fulfill other legitimate, and often necessary, functions.

More difficult are those cases in which reasons are indeed offered but more in way of clarification rather than justification. Although appearing like arguments, such passages are often no more than a collection of statements, one expanding on the other. Consider, for example, the following remark by nationally syndicated televangelist James Robinson:

d) Women have great strengths, but they are strengths to help the man. A woman's primary purpose in life and marriage is to help her husband succeed, to help him be all God wants him to be.

Robinson's main point is that women's role in life is to help the man, a point he then simply reaffirms and expands on in the rest of his comment. What we have then here is, essentially, merely a slightly elaborated but unsupported statement of an opinion—not an argument.

The same is true of the following oft-quoted aphorism of Francis Bacon:

> e)   He that hath wife and children hath given hostages to fortune; for they are impediments to great enterprise, either of virtue or mischief.

Rather than offering reasons why, in his view, women and children stand in a man's way (are *hostages to fortune*), Bacon simply explains himself by expanding and repeating the point. So this too is not an argument.

But cases will not always be so clear-cut and, often enough, one will meet examples where explanation and justification simply blend into each other:

> f)   We are sorry but we tried and tried but we find that the stains on this garment cannot be removed without possible injury to the color or fabric. This has been called to your attention so that you will know it has not been overlooked.

This passage can be said both to explain why the stains *have not* been removed as well as to offer a reason *why* they have not. Although the former is probably its main object, the explanation is of such a nature that it can function as a reason—i.e., constitute an argument—as well (and would, no doubt, be invoked as such were the need to arise).

It would seem best to evaluate examples like these in light of their primary intention. If, as in the case above, that intention is to explain rather than justify, then, strictly speaking, the passage is not an argument.

EXERCISES

Which of the following are arguments and which are nonarguments? If they are not arguments, explain what they are; if they are arguments, say why:

*12.   Back in the mid-1950s there was an actor named John Bromfield. I believe he was married to actress Corinne Calvet. I can't think of any movies that he was in, but he seemed quite popular for a year or so in TV. What happened to him?

13. Mother's Day Card:

CERTIFICATE OF AWARD

This is to certify that _____, having completed another year of marriage as a kind, lovable, and tolerant good sport in spite of much harassment and good-natured kidding, and having managed to come through the ordeal looking twice as sexy as ever, is hereby awarded the title Wife-of-the-Year, and is hereby authorized to enjoy the rights and privileges thereof (as soon as you can find the time). Signed this Mother's Day, _____ 19___ by _____

14. We must stop the homosexuals dead in their tracks—before they get one step further toward warping the minds of our youth. The time for us to attack is now! The enemy is in our camp! (Jerry Falwell, head of Moral Majority)

* 15. The main issue in life is not the victory but the fight. The essential thing is not to have won but to have fought well. (Baron Pierre de Coubertin, founder of the modern Olympic Games)

16. Sex deepens love and love deepens sex, so physical intimacy transforms everything and playing with it is playing with fire. Men try to ignore the fact that making love creates bonds, creating dependencies where there were none before, and women who try to ignore it with them deny their basic needs. (Merle Shain, *Some Men Are More Perfect Than Others*)

17. I must study politics and war, that my sons may have liberty to study mathematics and science. My sons ought to study mathematics and science, geography, natural history and engineering, commerce, and agriculture, in order to give their children a right to study painting, poetry, music, architecture, literature, and philosophy.

18.   Oh, come with old Khayyam, and leave the Wise
        To talk; one thing is certain, that life flies;
          One thing is certain, and the Rest is Lies;
        The Flower that once has blown for ever dies.

> Ah, make the most of what we yet may spend,
> Before we too into the Dust Descend;
>   Dust into Dust, and under Dust, to lie,
> Sans Wine, sans Song, sans Singer and sans End!

(Omar the Tentmaker, "Come With Old Khayyam")

* 19. You shouldn't have legislation against a thing that the majority of the population does. And today the majority smokes marijuana. So marijuana should be legalized.

20. It is important that you study this book thoroughly and with diligence. The state insurance departments which administer insurance examinations take seriously their responsibility to protect the public from unqualified persons. For that reason, life insurance examinations cannot be considered easy to pass. Prospective agents who take their task of studying lightly have been surprised to learn that they have failed examinations. Fortunately, those prospective agents who were qualified undertook study with more earnestness and passed examinations on subsequent occasions. While examinations cannot be considered easy, neither should they be considered unreasonably difficult. The purpose of the examination is to test your knowledge of the type of information contained in this book. If you study thoroughly, you have a good chance of passing. (Gary H. Snouffer, *Life Insurance Agent*)

21. The only creatures on earth that have bigger—and maybe better—brains than humans are the Cetacea, the whales and dolphins. Perhaps they could one day tell us something important, but it is unlikely that we will hear it. Because we are coldly, efficiently and economically killing them off. (Jacques Cousteau)

22. What Happens to Cigarette Smoke in the Air? The logical and obvious thing: cigarette smoke is immediately diluted by surrounding air. And measurements of cigarette smoke in the air, taken under realistic conditions, show again and again that there is minimal tobacco smoke in the air we breathe. In fact, based on one study, which measured nicotine in the air, it has been said that a nonsmoker would have to spend 100 straight hours in a smoke-filled room to consume the equivalent of a single filter-tipped cigarette. That's what we mean by minimal. So, does cigarette smoke endanger nonsmokers? In his most recent report, the sur-

geon general, no fan of smoking, said that the available evidence is not sufficient to conclude that other people's smoke causes disease in nonsmokers. In our view, smoking is an adult custom and the decision to smoke should be based on mature and informed individual freedom of choice.

23. Although we do not know if other influenza vaccines can cause Guillain-Barre Syndrome (GBS), this risk may be present for all of them. Little is known about the exact causes of GBS, but clearly the great majority of the several thousand GBS cases that occur each year in the United States are not due to influenza vaccine. The risk of GBS from the vaccine is very small. This risk should be balanced against the risk of influenza and its complications. The risk of death from influenza during a typical epidemic is more than 400 times the risk of dying from any possible complications of influenza vaccine injections.

4. ELIMINATING VERBIAGE

Arguments as ordinarily expressed are encumbered with a great deal of repetition, verbosity, and irrelevance. In order to see more clearly what such arguments are about it is necessary to scrape away a good deal of this deadwood from them. Sometimes this may simply involve ignoring a rather long, drawn-out introduction as in the following example from Og Mandino's classic sales manual *The Greatest Salesman in the World,* noted above:

a) Today I will be master of my emotions. The tides advance; the tides recede. Winter goes and summer comes. Summer wanes and the cold increases. The sun rises; the sun sets. The moon is full; the moon is black. The birds arrive; the birds depart. Flowers bloom; flowers fade. Seeds are sown; harvests are reaped. All nature is a circle of moods and I am part of nature and so, like the tides, my moods will rise; my moods will fall. Today I will be master of my emotions.

What this argument basically asserts is contained in its second-to-last sentence: "All nature is a circle of moods and I am part of nature and so, like the tides, my moods will rise; my moods will fall." For purposes of logical evaluation, all the preceding is irrelevant, however poetic and moving it may be.

What applies to introductions applies similarly to conclusions which may needlessly repeat what has already been adequately stated. The insurance example noted above is characteristic of such unnecessary repetition. Do the last three sentences of the passage say anything that has not been said previously?

> b) It is important that you study this book thoroughly and with diligence. The state insurance departments which administer insurance examinations take seriously their responsibility to protect the public from unqualified persons. For that reason, life insurance examinations cannot be considered easy to pass. Prospective agents who take their task of studying lightly have been surprised to learn that they have failed examinations. Fortunately, those prospective agents who were qualified undertook study with more earnestness and passed examinations on subsequent occasions. While examinations cannot be considered easy, neither should they be considered unreasonably difficult. The purpose of the examination is to test your knowledge of the type of information contained in this book. If you study thoroughly, you have a good chance of passing.

Often, however, an example is simply verbose throughout and needs to be abbreviated drastically before its structure can be observed clearly, as in the cigarette smoke example (no. 22) noted earlier. Before such arguments can be properly evaluated they need to be rewritten as concisely as possible and their premises and conclusion arranged in their logical order.

Revising the three arguments just noted along such lines, we find that they essentially state the following:

> a) All nature is a circle of moods; I am part of nature; therefore, I, too, must accept the fact that I will be subject to such swings of mood.
> b) Insurance examinations are not easy to pass without proper preparation. Therefore, to prepare yourself for them properly, buy and carefully read this book.
> c) According to a recent study, cigarette smoke is immediately diluted by surrounding air; this is also affirmed by the surgeon general; therefore, cigarette smoke in the air does not represent a serious danger to nonsmokers.

In eliminating such excess verbiage from arguments we will at times be forced to discard some of the "poetry" or literary elegance of the original. This is a sacrifice, however, we will need to make for the sake of logical clarity.

EXERCISES

Rewrite the following arguments as concisely as possible, clarifying their meaning and arranging the premises and conclusion in their logical order.

*24. I no longer believe those who say that a poor politician could be a good President, "if he could only be appointed to the job." Without the qualities required of a successful candidate—without the ability to rally support, to understand the public, to express its aspirations—without the organizational talent, the personal charm, and the physical stamina required to survive the primaries, the convention, and the election—no man would make a great President, however wise in other ways he might be. (Theodore G. Sorensen, *Decision-Making in the White House*)

*25. The day *may* come when the rest of the animal creation may acquire those rights which never could have been withholden from them but by the hand of tyranny. The French have already discovered that the blackness of the skin is no reason why a human being should be abandoned without redress to the caprice of a tormentor. It may one day come to be recognized that the number of the legs, the villosity of the skin, or the termination of the *os sacrum,* are reasons equally insufficient for abandoning a sensitive being to the same fate. What else is it that should trace the insuperable line? Is it the faculty of reason, or perhaps the faculty of discourse? But a full-grown horse or dog is beyond comparison a more rational, as well as a more conversable animal, than an infant of a day, or a week, or even a month, old. But suppose they were otherwise, what would it avail? The question is not, Can they *reason?* nor Can they *talk?* but, Can they *suffer?* (Jeremy Bentham, *The Principles of Morals and Legislation,* 1789)

26. Although we do not know if other influenza vaccines can cause Guillain-Barre Syndrome (GBS), this risk may be present for all of them. Little is known about the exact causes of GBS, but clearly the great majority of the several

thousand GBS cases that occur each year in the United States are not due to influenza vaccine. The risk of GBS from the vaccine is very small. This risk should be balanced against the risk of influenza and its complications. The risk of death from influenza during a typical epidemic is more than 400 times the risk of dying from any possible complications of influenza vaccine injections.

*27.   Oh, come with old Khayyam, and leave the Wise
        To talk; one thing is certain, that life flies;
            One thing is certain, and the Rest is Lies;
        The flower that once has blown for ever dies.

        Ah, make the most of what we yet may spend,
        Before we too into the Dust Descend;
            Dust into Dust, and under Dust, to lie,
        Sans Wine, sans Song, sans Singer and sans End!

        (Omar the Tentmaker, "Come With Old Khayyam")

28.   Because the father of poetry was right in denominating poetry an imitative art, these metaphysical poets will, without great wrong, lose their right to the name of poets, for they copied neither nature nor life. (Samuel Johnson, *Life of Cowley*)

29.   The Abbé, talking among friends had just said, "Do you know, ladies, my first penitent was a murderer," when a nobleman of the neighborhood entered the room and exclaimed, "You there, Abbé? Why, ladies, I was the Abbé's first penitent, and I promise you my confession astonished him." (Story by Thackeray)

30.   Surely also there is something strange in representing the man of perfect blessedness as a solitary or a recluse. Nobody would deliberately choose to have all the good things in the world, if there was a condition that he was to have them all by himself. Man is a social animal, and the need for company is in his blood. Therefore the happy man must have company, for he has everything that is naturally good, and it will not be denied that it is better to associate with friends than with strangers, with men of virtue than with the ordinary run of persons. We conclude then that the happy man needs friends. (Aristotle, *Ethics*)

31.   Forty years ago, it took farmers three to four months and five pounds of natural feed to produce one pound of

chicken meat. Today, it takes nine weeks and two and a half pounds of "doctored feed" to achieve the same results. The breeders are experimenting with techniques to do it with two pounds of feed. Today, 90 percent of all chickens eat arsanilic acid, an arsenic substance which is mixed into the feed as a growth stimulant. This substance is toxic to humans but apparently not to the chickens. To help chickens resist disease before they make it to the super-market, they are automatically given antibiotics. The F.D.A. also permits breeders to dip the slaughtered hens into an antibiotic solution which is designed to increase the shelf life of the chicken. Many other drugs and addi-tives are often added to poultry feed, among them tranquil-izers, aspirin, and hormones. How many of these chemi-cals come to affect us no one really knows.

* 32. Nothing in the world—indeed nothing even beyond the world—can possibly be conceived which could be called good without qualification except a *good will.* Intelligence, wit, judgment, and the other talents of the mind, however they may be named, or courage, resoluteness, and perse-verance as qualities of temperament, are doubtless in many respects good and desirable. But they can become extremely bad and harmful if the will, which is to make use of these gifts of nature and which in its special consti-tution is called character, is not good. It is the same with the gifts of fortune. Power, riches, honor, even health, gen-eral well-being, and the contentment with one's condition which is called happiness, make for pride and even arro-gance if there is not a good will to correct their influence on the mind and on its principles of action so as to make it universally conformable to its end. It need hardly be mentioned that the sight of a being adorned with no fea-ture of a pure and good will, yet enjoying uninterrupted prosperity, can never give pleasure to a rational impartial observer. Thus the good will seems to constitute the indis-pensable condition even of worthiness to be happy. (Kant, *Foundations of the Metaphysics of Morals*)

5. HIGHLIGHTING SUSPECT ELEMENTS

Among the things that are tempting to eliminate—once we come to recognize clearly the essential components of an argument—is a kind of verbiage that will occupy our attention in all of part two of this text. These are elements which, although they often arouse our suspicions, tend to leave us at a loss as to how to respond. We will examine them in great detail subsequently. However, because they are omnipresent and often influence us strongly, it is important that we become aware of them—if only in a very preliminary way—at this early point in our work on arguments.

In the context of argument, an element is suspect if it is made to carry more weight than it is capable of doing. To determine whether this is so in any given case we need to ask such questions as: are the facts cited in support of a conclusion sufficient as such a support; are they relevant; are they indeed so? Further, are the illustrations apt? Are they meaningful, and to what degree so? Is the wording clear so that we know for certain what its author has committed himself or herself to?

To illustrate:

a) In a press conference on May 13, 1982, President Reagan said the following regarding missiles carried by submarines and bombers: "You are dealing there with a conventional type of weapon or instrument, and those instruments can be intercepted. They can be recalled if there has been a miscalculation."

In using the term *they* here, did Reagan intend to refer to the missiles or the submarines? Could the missiles be recalled? Did Reagan somehow believe at that time they could? His critics claimed he did and they referred to the remark as their proof of this; Reagan claimed he was referring to the submarines and referred to the same remark as *his* proof.

b) When on September 27, 1984, a terrorist zigzagged an explosive-laden truck around concrete barriers and set off a blast that once again killed many people at the still unfinished and poorly secured American embassy in Beirut, President Reagan replied as follows to critics: "Anyone that's ever had their kitchen done over knows that it never gets done as soon as you wish it would."

Is this an apt illustration? Is the analogy meaningful here? Many people were naturally struck by the inaptness of Reagan's illustration—for, after all, a commander in chief presumably has a good deal more influence over his forces than we have over contractors, to say nothing of the differences in the consequences.

We need not be masters of the art of reasoning to know our rights in argument, and we should not, therefore, hesitate to demand proof when a presentation or an account seems overblown or questionable. For example, in the following argument noted earlier:

c) You shouldn't have legislation against a thing that the majority of the population does. And today the majority smokes marijuana. So marijuana should be legalized.

Is it really the case that we "shouldn't have legislation against a thing that the majority of the population does"? Does this apply in the case of a majority being, say, prejudiced against another group within its midst? Should we not make an effort to change the will of such a majority?

And in the cigarette smoke example, to refer to it once more, is that one study mentioned really sufficient? And who conducted the study; under whose auspices? The surgeon general, we are told further, said that the available evidence is not sufficient to conclude that other people's smoke causes disease in nonsmokers. That is a very cautious remark from which it would be fallacious to infer—if the passage suggests we do—that such smoke is *not* such a cause or factor. It may very well still be such a cause or factor.

EXERCISES

In the statements and arguments that follow, underline and explain any suspect elements—illogical or ambiguous phrases, inapt illustrations, half-truths, faulty analogies, etc.—that may be important in determining and evaluating their meaning and soundness.

33. Clean and Decent Dancing Every Night Except Sunday (Roadhouse sign)

*34. If you don't go to other people's funerals, they won't come to yours.

35. If you can tie a knot, you can make a beautiful deep pile rug! (Advertisement)

* 36. The end of a thing is its perfection; death is the end of life; death is, therefore, the perfection of life.

37. Abraham Lincoln was obviously not the man it is thought he was, for he said, "You can fool all the people some of the time."

38. Doctors are all alike. They really don't know any more than you or I do. This is the third case of faulty diagnosis I've heard of in the last month.

* 39. On November 5, three of the accused met at the house of the fourth defendant, Smith. There, behind locked door and heavily curtained windows, these four conspirators began to hatch their dastardly plot.

40. Raising children, after all, requires skills that are by no means universal. We don't let "just anyone" perform brain surgery or, for that matter, sell stocks and bonds. Even the lowest ranking civil servant is required to pass tests proving competence. Yet we allow virtually anyone, almost without regard for mental or moral qualifications, to try his or her hand at raising young human beings, so long as these humans are biological offspring. Despite the increasing complexity of the task, parenthood remains the greatest single preserve of the amateur. (Alvin Toffler, *Future Shock*)

41. The meat cutters and the retail clerks merged in 1979, becoming one of the largest labor unions in the United States. This growth is like the sunset. The sun gets largest just before it goes under. Remember the brontosaurus? The brontosaurus got so huge just before its demise that it had to stay in the water to remain upright. There were thirty-five mergers of labor unions between 1971 and 1981. In late 1980, the machinists and the auto workers announced merger talks that would make them the largest, most powerful union in the United States. If it ever comes off, it will appear that big labor is getting its act together again. The reality will be the sunset effect. (John Naisbitt, *Megatrends*)

* 42. You say you're a vegetarian for moral reasons but no one has proven yet that eating meat is unhealthy.

43. My gun has protected me, and my son's gun taught him safety and responsibility long before he got hold of a far

more lethal weapon—the family car. Cigarettes kill many
times more people yearly than guns and, unlike guns, have
absolutely no redeeming qualities. (Letter to the Editor
opposing gun-control)

6. MISSING COMPONENTS

An argument's basic structure, as we saw, may be obscured by an
excess of verbiage, an excess which we need to get rid of so that
premises and conclusion stand out clearly before us. But an argument's structure may also be obscured for us (and as a result
possibly misleading and deceptive) because it is too sparse and has
missing components. Such arguments may appear sounder than
they are because we are unaware of important assumptions made
by them and on which they rest. Such assumptions need to be dug
out, if hidden, or made explicit, if unexpressed. Once they are
made explicit, it will be easier to determine the role these missing
components (assumptions) play in the argument and to what degree the argument depends upon them.

It will be easier to find such missing components of an argument
if we will keep in mind that many arguments consist of a statement of a general principle, the citing of a case of it, and a conclusion which infers that what is true of the general principle is true
of the case in question. The following is a classic example:

a) All men are mortal. (the general principle)
    Socrates is a man. (the case)
    Socrates is mortal. (the conclusion)

In the exercises above, the examples from Aristotle (no. 30) and
Samuel Johnson (no. 28) conform very much to this pattern. In
abbreviating and putting them into proper logical form, what you
got was probably the following:

b) To be happy is to have the things you need.
    One of the things you need is friends.
    To be happy, therefore, you need friends.
c) Poetry is an imitative art.
    Metaphysical poetry is not imitative.
    Metaphysical poetry, therefore, is not poetry.

Arguments of this type may therefore lack either the statement of the general principle (called in logic the major premise), explicit reference to the case in question (the minor premise), or the inference (the conclusion). Some typical examples:

d) These are natural foods and therefore good for you.

Omitted here is the major premise: all natural foods are good for you:

→ All natural foods are good for you.*
These are natural foods.
These foods are good for you.

e) You'll make an excellent kindergarten teacher. People who are fond of children always do, you know.

Omitted here is the minor premise: you are fond of children:

All who are fond of children make excellent teachers.
→ You are fond of children.
You'll make an excellent kindergarten teacher.

f) "Yon Cassius has a lean and hungry look; such men are dangerous."

Omitted here is the conclusion: Cassius is dangerous:

All who have lean and hungry looks are dangerous.
Cassius has such a look.
→ Cassius is dangerous.

Not all such omissions are innocent, or done for the sake of literary elegance or brevity. Often what is omitted is highly questionable and omitted for that very reason:

*The arrow indicates the missing component.

g) This must be a good book; it was chosen by the Book-of-the-Month Club.

What has been left unstated here is the major premise: All books chosen by the Book-of-the-Month Club are good. For understandable reasons, to state it explicitly is to call attention to it and risk having it questioned. The same is true of the following two examples:

h) All alcoholics are short-lived; therefore Jim won't live long.
i) Cowardice is always contemptible, and this was clearly a case of cowardice.

Although the missing components easily spring to mind in such brief examples, it is still an advantage not to state them explicitly, for to do so is, again, to call attention to them and risk a challenge.

More difficult to unravel, and far more frequent, are the longer and more verbose examples. More turns on them—the opportunity for gain, influence, deception—and hence a greater effort is made to hide the assumptions on which the argument rests. Because the arguments are more complicated verbally, it is easy to lose track of the missing components. The following is an advertisement for a tape dealing with loneliness:

j) Almost everyone feels lonely at times, and for many it is a constant companion. From the little child who feels he has no friends to the elderly who feel resigned to a cold empty feeling. Teenagers often feel they are nobody and young adults feel friendless. This remarkable tape is not only comforting news but also contains an innovative approach to resolving this emptiness.

Restating this argument, we find that it asserts the following:

Everyone suffers from loneliness.
This tape is a cure for loneliness.
This tape will relieve *your* loneliness.

It is the conclusion that we now see has been omitted—and for good reason. To state it explicitly possibly raises a question in one's mind: Even if it is true that this tape has helped others, will it help me?

"So, you're a *real* gorilla, are you?
Well, guess you wouldn't mind munchin' down
a few beetle grubs, would you? ... In fact,
we wanna see you chug 'em!"

What are the missing components of the argument? The visitor is either a human or a gorilla. All gorillas, the speaker implies, love grubs. If the visitor is a gorilla, he will eat the grubs. Humans, however, do not like to eat grubs. If the visitor declines the offer, then he is human.

Not all omissions of this sort are, of course, questionable. Usually a person will omit a component because it is simply too obvious to state explicitly, sometimes one does so for dramatic reasons, and occasionally because one wishes to be somewhat guarded and cautious. Two of the examples in the exercises were of this last sort: the Guillain-Barre Syndrome (GBS) example (no. 26) and the poultry example (no. 31). Both writers seemed hesitant to state their conclusions. Highly abbreviated, in the GBS announcement we were told that the risk of death from influenza during a typical epidemic is far greater than the risk of dying from the complications of influenza vaccine injection, and it was left to us to infer that, this being so, vaccination was therefore the wiser choice. In the poultry example, we were informed that to stimulate growth and increase shelf life, poultry is now fed and treated with large doses of a variety of potentially dangerous chemicals. The conclusion that, this being so, it would be wiser to investigate the risks to us of these new breeding and marketing methods was, again, merely implied. It was not explicitly stated.

Such hesitancy, if that is all it is, may be defensible from the point of view of scientific reserve or legal caution; in logic, however, we always make it a point to know clearly what is being asserted and what we are being asked to assent to.

EXERCISES

Supply the missing components in the following arguments. If these components, required by the argument, throw the argument into suspicion explain how and why so.

* 44. The speaker criticizes free enterprise; he must be a communist.

45. She is a Phi Beta Kappa so she must have been a bookworm.

46. The energy crisis, being man-made, can be man-solved.

47. Our ideas reach no farther than our experience; we have no experience of divine attributes and operations; I need not conclude my syllogism; you can draw the inference yourself. (David Hume)

* 48. We do not want a democracy in this land, because if we have a democracy a majority rules. (Televangelist Rev. Charles Staney)

49. It is said that a school is only as good as its teachers. We at Melrose Street Nursery School have one of the finest staffs in the city.

50. Death cannot be an evil, being universal. (Goethe)

51. No person is free, for every person is a slave either to money or to fortune.

52. Pleasure is good, since it is sought by everyone.

* 53. Nothing intelligible puzzles me, but logic puzzles me.

7. DEDUCTIVE AND INDUCTIVE ARGUMENTS

Having distinguished between arguments and nonarguments, separated premises from conclusions, eliminated excess verbiage, highlighted for later analysis suspect elements, and supplied missing components, it remains for us to ask two important and crucial questions of an argument: Are the premises true? Does the conclusion really follow from them?

Regarding the first question we want to know whether the facts stated by the argument are really so. Or do they perhaps misrepresent or falsify them? Do they prejudge them? Are they, perhaps misleading as stated? Premises, after all, are the foundation of an argument; if they are unreliable or shaky, the argument built on them will be no better.

There is, however, another way an argument can go wrong: when the relationship between the premises and conclusion is such that the premises fail as a support of the conclusion in question.

A premise can support a conclusion either:

Fully: All men are mortal.
                Socrates is a man.
                Socrates is mortal.

Partially: Most Scandinavians are blonde.
                My cousin Christine is Scandinavian.
                She is blonde too.

                Or

Not at all: "Be sure to brush with Colgate.
                Walt Frazier wouldn't think of
                brushing with anything else."

We will consider arguments of the third type—seductive arguments, let us call them—in great detail in part two. Here let us consider the first two: the first is called deductive, the second inductive. Deductive arguments are arguments in which the conclusion is presented as following from the premises *with necessity.* Inductive arguments, on the other hand, are arguments in which the conclusion is presented as following from the premises only *with probability.*

Two examples will help illustrate this distinction between necessary and probable inference.

a) *Deductive*

All the pears in that basket are ripe.
All these pears are from that basket.
All these pears are therefore ripe.

b) *Inductive*

All these pears are from that basket.
All these pears are ripe.
All the pears in that basket are therefore ripe.

Of these two arguments, only the first (argument *a*) has a conclusion that follows with certainty from its premises; the conclusion of argument *b* follows only with some degree of probability from its premises.

One difference between deductive and inductive arguments, it will be observed, is that the premises in a deductive argument contain all the information needed in order to reach a conclusion that follows with necessity. Nothing in the conclusion refers outside the premises. In the conclusion of an inductive argument, on the other hand, we must venture beyond information contained in the premises. Thus our conclusion can never be certain, although it can have a high probability of being true.

It is because deductive arguments either prove or fail to prove their conclusions with certainty that we say of them that they are either valid or invalid; inductive arguments, on the other hand, are said to be either good or bad, strong or weak.

A classic example of inductive argument highlights this issue of certainty.

c) The sun has risen every morning since time immemorial.
Therefore the sun will rise tomorrow morning.

We feel sure that the sun will rise tomorrow, yet logically speaking the relation of this conclusion to its premises is one of probability, not necessity. (As the renowned logician Bertrand Russell once

"Coincidence, ladies and gentlemen?
Coincidence that my client just *happened*
to live across from the A-1 Mask Co., just
happened to walk by their office windows
each day, and they, in turn, just *happened*
to stumble across this new design?"

*The above argument is based on probability. One can't know,
with absolute certainty, from what source the A-1 Mask Co. got
their inspiration for the mask. However, the facts collected by the
attorney make for a strong inductive case on the behalf of his
client.*

put it, in *The Problems of Philosophy,* "The man who has fed the
chicken every day throughout its life at last wrings its neck instead.") In inductive arguments, we assert in the conclusion a fact
not itself contained in the premises. In argument *c* above, for
example, the premises make assertions only about the past; they
assert nothing about what will happen in the future. Therefore
the premises do not rule out the possibility of the conclusion being
false, since they yield a conclusion whose truth is only probable

with respect to these premises, not necessary. It is in the nature of inductive arguments to carry us beyond what is asserted in the premises so that we may see what implications those premises have for other events.

Deductive reasoning is precisely the reverse. Here the premises contain all the information that we seek to draw out or unfold. We attempt not to go beyond the premises but to understand more specifically what they contain. In the following example, everything contained in the conclusion is contained, either explicitly or implicitly, in the premises:

> d) If there are 50,001 people in a town,
> And if no person can have more than 50,000 hairs on his or her head,
> And if no one is completely bald,
> Then at least two people in the town have the same number of hairs on their heads.

This example illustrates the precision of which deduction is capable. Whereas inductive arguments expand the content of their premises at the sacrifice of necessity, deductive arguments achieve necessity by sacrificing expansion of content. Most of the arguments one encounters in daily affairs are of the inductive type, and it is these that we shall treat most extensively in this book.

These types of arguments, however, are alike in having premises and a conclusion and hence must both be evaluated in light of our two basic questions: one, are the premises true, and, two, does the conclusion follow from them?

We will turn to this topic in the next section.

EXERCISES

Determine whether each of the following arguments is deductive or inductive. Give your reasons for your answer in each case.

*54. There are no foxes in this area. We haven't seen one all day.

55. Tom must have left already. He doesn't answer his phone.

56. "How, in the name of good fortune, did you know all that, Mr. Holmes?" he asked. "How did you know, for example, that I did manual labor? It's true as gospel, for I began as a ship's carpenter."

    "Your hands, my dear sir. Your right hand is quite a size larger than your left. You have worked with it and the muscles are developed." (Arthur Conan Doyle, "The Red-headed League")

57. Because many drug addicts who came through the courts admit that they started on marijuana, marijuana likely causes addiction to hard drugs.

58. Because the father of poetry was right in denominating poetry an imitative art, these metaphysical poets will, without great wrong, lose their right to the name of poets, for they copied neither nature nor life. (Samuel Johnson, *Life of Cowley*)

*59. Everyone in the chemistry class needed to have had one year of high-school chemistry as a prerequisite. Since John is a member of that class, he must have had one year of high-school chemistry.

60. The house across the street has shown no signs of life for several days. Some rain-soaked newspapers lie on the front steps. The grass badly needs cutting. The people across the street therefore must be away on a trip.

*61. All life requires water. There is no water on the planet Venus. Therefore there is no life on that planet.

62. Tom will be ineligible to vote in this state, for he is nineteen years old and only persons over the age of twenty-one are eligible to vote here.

63. No Athenian ever drank to excess, and Alcibiades was an Athenian. Therefore, Alcibiades never drank to excess.

*64. Every event in the world is caused by other events. Human actions and decisions are events in the world. Therefore, every human action and decision is caused by other events.

65. Our ideas reach no farther than our experience; we have no experience of divine attributes and operations; I need not conclude my syllogism; you can draw the inference yourself. (David Hume)

8. EVALUATING ARGUMENTS: TRUTH, VALIDITY, AND SOUNDNESS

People are sometimes heard to say, "That may be logical, but it's not true," or, "What's logical isn't always right." Both of these views are correct, yet they do not mean that logic is unconcerned with *truth*. Indeed, logic defines truth rigorously and separates it from two other concepts—*validity* and *soundness*—with which it is sometimes confused in ordinary speech. Together, these three concepts provide a basis for evaluating any argument.

Validity refers to the correctness with which a conclusion has been inferred from its premises, whether the conclusion *follows* from them. Truth, on the other hand, refers to whether those premises and conclusion accord with the facts. It is thus possible in logic to start with true premises but reach a false conclusion (because we reason badly with those premises) or to reason correctly or validly without reaching a true conclusion (because our premises are false). Soundness results when the premises of an argument are true and its conclusion validly derived from them. Otherwise the argument is *unsound*.

In order to accept the conclusion of an argument as sound, therefore, we must be sure of two things. We need to know, first, that the premises are true, not false. Premises, after all, are the foundation of an argument; if they are unreliable or shaky, the argument built on them will be no better. Second, we need to know that the inference from the premises is valid, not invalid—that the conclusion *follows* from the premises. One may begin with true premises but make improper use of them, reasoning incorrectly and thus reaching an unwarranted conclusion.

Truth and falsity, validity and invalidity, can appear in various combinations in argument, giving rise to these four possibilities:

1. We may have our facts right (our premises are true), and we may use them properly (our inference is valid). In such a case not only will our argument be valid but our conclusion true. The argument as a whole will be sound.

    a) All men are mortal.
    Socrates is a man.
    Therefore Socrates is mortal.

2. We may have our facts right (our premises are true), but we may make improper use of them (reason invalidly from them). In this case our conclusion will not follow, and the argument as a whole will be unsound.

b) All cats are animals.
   All pigs are animals.
   Therefore all pigs are cats.

On some occasions the conclusion of such an argument may accidentally happen to be true, as in:

c) All cats are animals.
   All tigers are animals.
   Therefore all tigers are cats.

In such a case we cannot determine the truth of the conclusion from the argument itself; the conclusion may be true but not for the grounds offered in defense of it in this argument.

3. We may have our facts wrong (one or more of our premises is false), but we may make proper use of them (reason validly with them). In this case, our argument will be valid but unsound.

d) All movie stars live in Hollywood.
   Robert Redford is a movie star.
   Therefore Robert Redford lives in Hollywood.

Here the first statement is clearly false, yet the reasoning is valid and the conclusion follows from the premises. As in case 2 above, the conclusion may happen to be true but we cannot determine its truth within the terms of the argument. It might be true despite the falsity of the first premise; on the other hand, it might be false despite the validity of the reasoning. In order to reach a conclusion that we can depend on to be true, it is not enough to reason validly; we must do so from true premises.

4. There is, finally, the case in which our facts are wrong (one or more of our premises is false) and we also make improper use of them (reason invalidly from them). In such a case the argument will be both invalid and unsound.

e) I like this course.
   All final examinations are easy.
   Therefore I will receive a high grade in this course.

We can summarize the points covered in this analysis of truth, validity, and soundness in arguments as follows:

1. Truth and falsity are descriptive of the properties of statements alone.
2. Validity and invalidity refer to reasoning and are determined independently of the truth or falsity of the premises or conclusion of the argument.
3. If an argument is valid and its premises are true, the conclusion either *must be true*—if it is a deductive argument—or *will probably be true*—if it is an inductive argument.
4. If in addition to being valid an argument contains true premises, the argument must be considered sound. Otherwise, it must be considered unsound. All sound arguments, therefore, are valid, but valid arguments can be either sound or unsound.

Table 1.1 summarizes these relations.

Table 1.1   The Four Types of Argument

	Premises	Validity	Soundness
1	T	V	S
2	T	I	U
3	F	V	U
4	F	I	U

Since only one of the argument types we have discussed can yield conclusions that must be true, the reader may wonder why we should be interested in arguments whose premises are false. For better or worse, we are sometimes in a position where we do not know whether our premises are true. Being able to infer validly the consequences which would flow from such premises if they were true enables us to judge whether they are true. For if, by a deductively valid inference, we should arrive at a conclusion that we know is false, then we can be sure that at least one of our premises is false, because a false conclusion cannot validly be deduced from true premises. An interesting example from the history of science concerns the formerly held corpuscular theory of light. This theory maintained that particles of light must travel in straight lines through empty space, but it eventually was realized that if this theory were true, then light particles traveling through a circular hole in an opaque shield would project a sharply defined circle of light onto a screen behind the shield. In a subsequent experiment using a very tiny hole, however, the image projected on the screen was not a sharply defined circle of light at all, but rather consisted of concentric alternating light and dark rings. The experiment showed that light does not travel in straight lines but rather in wavelike undulations. The corpuscular theory came to be replaced with the wave theory of light.

Knowing, therefore, that something can follow from something else even though what it follows from is false can be enormously useful. For this means that if you are uncomfortable with a conclusion seemingly validly derived from a premise, it is possible you are not in full agreement with the premise from which it is, apparently, correctly deduced. The trouble may therefore lie in the premise.

Consider, for example, the following argument:

> f) Abortion is the destruction of a human fetus, and the destruction of a human fetus is the taking of a human life. If, therefore, the taking of a human life is murder, then so is abortion.

What are the premises of this argument? What is the conclusion? Does the conclusion follow validly from the premises? How would you challenge this argument?

The form of sound deductive arguments is equally useful. For if we reason validly from true premises, we must necessarily arrive at a conclusion that is true whether or not we can test its truth directly.

EXERCISES

Which of the following statements are true and which are false?

*66.   If the conclusions of our arguments have been derived validly from our premises, then we know that such conclusions are true.

67.   If the conclusion of our argument has been derived validly from our premises, then we know that this argument is sound.

68.   If our premises are true, and our conclusion has been validly derived from these premises, such an argument is called sound.

69.   All valid arguments are sound.

*70.   All sound arguments are valid.

71.   A conclusion validly derived from premises may nevertheless be false.

*72.   A conclusion invalidly derived from the premises may nevertheless be true.

73.   Deductive arguments are either valid or invalid; inductive arguments are either strong or weak.

74.   In a valid deductive argument, if the premises are true, the conclusion must be true.

75.   In a strong inductive argument, if the premises are true, the conclusion is probably true.

SUMMARY

This chapter has explored the kind of discipline logic is and what some of its elements are. We saw that, although it is a theoretical discipline, logic is at the same time a practical field of study, offering significant benefits for the conduct of everyday affairs.

We noted that logic is the study of argument and that every argument consists of two basic elements: premises and conclusion. This, we saw, is what distinguishes arguments from nonarguments. Not all arguments, however, display this structure so simply. As ordinarily expressed, arguments are encumbered with excessive verbiage and irrelevance, and often rest on hidden or unexpressed assumptions. We learned how to recognize and eliminate such excessive verbiage, how to expose these suspect elements, and how to identify hidden and unstated assumptions.

Going on with our study of argument, we saw further how an argument possesses three characteristics which are bases for evaluating it. The first of these is the truth or falsity of the premises. The second is the validity or invalidity of the reasoning from the premises. And the third characteristic is soundness, which exists whenever the premises are true and the reasoning valid, or unsoundness, which exists if truth and/or validity is lacking.

We then saw how in deductively valid arguments, the premises contain all the information necessary for the conclusion; while in inductive arguments the conclusion goes beyond the data contained in the premises. Thus even in the best inductive arguments, the conclusion is only probable, whereas the conclusion in a deductively valid argument follows with necessity.

One of the important lessons we learned here was that something (a conclusion) may indeed follow from something else (a premise) without the conclusion necessarily being true. Whether or not the conclusion is true will depend on whether the premise that it follows from is itself true. In short, a conclusion may follow from a premise, yet be false.

ANSWERS TO STARRED EXERCISES

1. Premise 1: All rational beings are responsible for their actions.
   Premise 2: All human beings are rational.
   Conclusion: All human beings are responsible for their actions.

5. Premise 1: The city should reimburse for hospital expenses anyone injured in an accident which takes place while he is engaged on city business.
   Premise 2: He was injured in an accident which took place while he was engaged on city business.
   Conclusion: The city should reimburse him for his hospital expenses.
   Premise 1 is unstated but clearly implied in the text.

9. Premise: Tom doesn't answer his phone.
   Conclusion: He must have left already.

12. An inquiry, not an argument.

15. Not an argument. The passage consists of two statements, the second elaborating on the first.

19.   This is an argument, consisting of a conclusion (that marijuana should be legalized) and two premises (the rest).

24.   To be a great president requires organizational talent, personal charm, physical stamina, etc.
       Good politicians have these qualities.
       A good politician can make a good president.

25.   Animals are similar to humans in being sensitive creatures, susceptible to suffering.
       This is a crucial similarity.
       Basic rights enjoyed by humans should, therefore, not be denied to them.

27.   Life passes one by very quickly into an endless death.
       Make the most, therefore, of what is yet left.

32.   Intelligence, wit, judgment, courage, etc. are good but not without qualification.
       They can become bad if combined with a bad will.
       This leaves only a good will itself as the only good thing without qualification in the world.

34.   If you don't go to other people's funerals, *they* won't come to yours. (*They* seems to refer to the people who died. This needs to be clarified so that it refers to the *relatives* of those who died.)

36.   The *end* of a thing is its perfection; death is the *end* of life; death is, therefore, the perfection of life. (The word *end* changes its meaning drastically in the course of this argument: at the beginning it means goal or purpose; the second time it is used in the sense of ending or termination— an entirely different thing.)

39.   On November 5, three of the accused met at the house of the fourth defendant, Smith. There, behind *locked* door and *heavily curtained* windows, these four *conspirators* began to *hatch* their *dastardly* plot. (Since persons alluded to in this passage are still only defendants, it is to prejudge the people and events described to refer to them with the language used here.)

42.   You say you're a vegetarian for *moral* reasons but no one has proven yet that eating meat is *unhealthy.* The person claims to be a vegetarian for moral reasons, not health reasons; whether eating meat is healthy or unhealthy is thus irrelevant.

44. *Major premise:* All who criticize free enterprise are communists. (An obviously questionable premise, not allowing for the possibility of loyal opposition.)

48. *Minor premise:* We don't want the majority to rule. (Rev. Staney may not want this but the rest of us may feel differently about this.)

53. *Conclusion:* Logic is not intelligible. (This does follow from the premises. Our recourse, then, as we will learn subsequently, is to question the truth of the premise or premises: see section 8.)

54. *Inductive.* The fact they haven't seen any foxes all day makes it only probable, not certain, that there aren't any in the area. (They may not be very observant or the foxes may be avoiding them.)

59. *Deductive.* Given that the premises are true, the conclusion follows from them with certainty: if everyone in that class absolutely had to have had one year of high-school chemistry in order to get into that class, and he is a member of it, then he must have had one year of high-school chemistry.

61. *Deductive.* If it is the case that all life—meaning all organic life—requires water and there is no water on Venus, then it follows necessarily that there is no life on that planet.

64. *Deductive.* If every event is caused and human action and decisions are events, then they too must be caused. So this conclusion also follows with necessity from its premises. (But, we might ask, is it the case that human actions and decisions are like ordinary events of the kind implied in the major premise?)

66. *False.* The premise from which it is derived may be false and therefore make the conclusion false too.

70. *True.* This statement is true by virtue of the definition of a sound argument. A *sound argument* means "a *valid* argument with all of its premises true."

72. *True.* It may accidentally turn out to be so.

"The Far Side" cartoon by Gary Larson is reprinted courtesy of Chronicle Features, San Francisco.

2

The Medium of Language

One important element in any argument was touched on only briefly in chapter 1. That element is the actual language in which the premises and conclusion are expressed. Because the words used in argument are so crucial, it will be useful to examine here certain features of language that can aid or hinder clear thinking. We shall explore the relationship between language and thought and will illustrate some of the confusions that result from the use of imprecise language. We shall see that all language is symbolic and that language usage lies at the heart of many disputes.

When we are confronted with an argument, we need to know whether it is clear. We must agree upon the meanings of all its words and phrases, singly and in combination. Because language is our tool of verbal communication, the study of logic must concern itself with the structure and function of language. Here we examine the medium of language and try to discover why some arguments succeed in communicating their meaning to us while others do not.

1. LANGUAGE AND THOUGHT

"Men imagine that their minds have the command of language, but it often happens that language bears rule over their minds." So remarked Francis Bacon, a philosopher deeply committed to clarity of observation and of thought. Bacon's statement is useful in reminding us that language can hamper not only communication of thought but thought itself. Because of the gradual, natural way in which we acquire language, we seldom pause to observe that it is an instrument and that, like all instruments, its value depends on the skill of the person using it.

The relationship between language and thought is an age-old question. In the past, two views were dominant: one holding that language is merely the vehicle or outer garment of thought; the other maintaining that the stream of language and the stream of thought are one, that thought is merely soundless speech.

More recently, research has tended to confirm the view that language and thought are intimately connected, that language is not merely sound but a union of sound and sense in which each is highly dependent on the other. Modern theories hold that words without thoughts are indistinguishable from other sounds to be found in nature. Such theories maintain, however, that, although we may have "vague thoughts" or ideas which we are unable to put into words, we cannot have a *clear* thought without being able to express it in language.

One who has made us especially aware of this linguistic condition of ours, although he has had his critics, is American linguist Benjamin Lee Whorf, who notes:

> When linguists became able to examine, critically and scientifically, a large number of languages of widely different patterns, their base of reference was expanded; they experienced an interruption of phenomena hitherto held universal, and a whole new order of significances came into their ken. It was found that the background linguistic system (in other words, the grammar) of each language is not merely a reproducing instrument for voicing ideas but rather is itself the shaper of ideas. . . . We cut nature up, organize it into concepts, ascribe significances as we do, largely because we are parties to an agreement to organize it in this way—an

agreement that holds throughout our speech community and is codified in the patterns of our language. (*Language, Thought and Reality,* ed. John B. Carroll. Cambridge, Mass.: M.I.T. Press, 1964, pp. 212–214)

According to this view, thought comes to be shaped by the language in which it is conducted. There are many examples which seem to support this hypothesis. The Zulus have words for "white cow" and "red cow" but no word for "cow." Lacking the word, they lack the idea as well. Similarly, the aborigines of central Brazil have no words that mean "palm" or "parrot," although they have a great number of concrete names for different varieties of palms and parrots. From such linguistic differences, we make assumptions about differing modes of thought. It is important, however, to try to avoid value judgments in all such comparisons. We sometimes tend to assume that the languages of so-called primitive peoples are as primitive as we suppose their lives to be. But this is true of neither their lives nor their languages. All languages, even those of the most primitive tribes, are already ancient. None of them is uncomplicated. Some are of a very high order of sophistication and make use of linguistic devices which, were our own language to incorporate them, would improve it. Certain constructions of the Zulus or the central Brazilians enable them to avoid some of the pitfalls that a language containing a greater number of abstractions may generate for its users.

Another and more practical view of the relation between language and thought is that of the British novelist and journalist George Orwell. In his famous essay "Politics and the English Language," Orwell argues that clear thought about the political realities of our time is difficult because our language has been corrupted by the rhetoric of politicians, who, in order to gloss over their brutal and untenable policies, regularly resort to euphemism and jargon (see sections 3 and 8).

Now, it is clear that the decline of a language must ultimately have political and economic causes: it is not due simply to the bad influence of this or that individual writer. But an effect can become a cause, reinforcing the original cause and producing the same effect in an intensified form, and so on indefinitely. A man may take to drink because he feels himself to be a failure, and then fail all the more completely because he drinks. It is rather the same thing that is happening to the English language. It becomes ugly and inaccurate because our thoughts are foolish, but the slovenliness of our language makes it easier for us to have foolish thoughts. (In *A Collection of Essays.* New York: Doubleday & Co., 1953)

But, Orwell goes on to argue, we need not resign ourselves to this condition.

The point is that the process is reversible. Modern English, especially written English, is full of bad habits which spread by imitation and which can be avoided if one is willing to take the necessary trouble. If one gets rid of these habits, one can think more clearly, and to think clearly is a necessary first step towards political regeneration: so that the fight against bad English is not frivolous and is not the exclusive concern of professional writers. (Ibid.)

Orwell has taught us that self-consciousness about the language we speak and the way we speak it is not merely a trivial interest of pedantic scholars and fussy school teachers. It is, or at least should be, a vital interest of everyone.

2. SIGNS AND SYMBOLS

Among the characteristics of language that contribute to its power as an instrument of communication is its *representational* quality, its use of words to stand for something beyond themselves. In order to appreciate this quality of language, we need to examine *signs* and *symbols* and to distinguish one from the other.

A sign is anything we use to refer to—or take as an indication of—something else. We speak, for example, of "the signs of life," or we say that smoke is a sign of fire. In such cases, the signs in question are *natural* signs. By a natural sign we mean that the connection between the sign and the thing it signifies actually exists in nature. The connection is therefore natural because it is not of our own making. We discover such signs; we do not invent them. They are parts or symptoms of the things they signify, as smoke is a sign of fire not because it represents fire but because it is part of the process of fire. When we refer to any sign of a natural phenomenon—such as a rash on the body of someone ill with measles—we say, not that it is a sign *for* the thing in question, but a sign *of* it.

A symbol, by contrast, is invented to stand for something. Symbols are therefore *conventional,* not natural, signs, for they are the conscious product of human thought and have no natural connection with what they represent. They are signs *for,* not signs *of.* Thus, if a weather station raises a flag of a certain color to warn the public of an approaching storm, this symbol will be effective only if observers understand its symbolic meaning. Because all symbols are artificial, it is we, not nature, who are the source of whatever meaning they have for us. This is true of nonlinguistic symbols such as the cross, the flag, and a red light. It is also true of linguistic symbols exhibiting *onomatopoeia*—that is, containing sounds that suggest their meanings, as do the sounds in *buzz, hiss, splash,* and *purr.* And it is ultimately true of all the words in our language. The word *month,* for example, carries no meaning unless we have learned that it symbolizes a certain period of time.

Without a system of arbitrary, nonimitative symbols, only the most limited communication would be possible. Most phenomena do not have characteristic sounds or sounds that the human voice can easily imitate. This no doubt explains why the number of onomatopoeic words in any language is small. Another reason why a language composed solely of imitative sounds would not carry us very far is that there is a vast body of ideas that are incapable of being represented in this way because they produce no sounds. This is true of most words in our language which stand for abstractions, such as *month.* Language succeeds in labeling such abstractions by making use of arbitrary, nonrepresentational symbols. Language, therefore, is essentially symbolic.

EXERCISES

Some of the following observations about language are famous; others are less well known. Which of them would you be inclined to agree with, and why? With which would you disagree? Why?

 * 1. No two languages can say quite the same thing. (Michael Roberts)

   2. In the last analysis we are governed either through talk or through force. (Felix Frankfurter)

   3. Spanish is the language for lovers, Italian for singers, French for diplomats, German for horses, and English for geese. (Spanish proverb)

 * 4. "The question is," said Alice, "whether you *can* make words mean so many different things." "The question is," said Humpty Dumpty, "which is to be master—that's all." (Lewis Carroll, *Alice in Wonderland*)

   5. Human language is like a cracked kettle on which we beat out tunes for bears to dance to, when all the time we are longing to move the stars to pity. (Gustave Flaubert)

   6. Most of us—almost all—must take in and give out language as we do breath, and we had better consider the seriousness of language pollution as second only to air pollution. (John Simon)

   7. All languages are the product of the same instrument, namely, the human brain. (David Foster, *A Primer for Writing Teachers*)

   8. How do I know what I think until I see what I say. (E. M. Forster)

3. WORDS AND THINGS

The distinction drawn above between signs and symbols stresses that, since all words function as symbols, no natural connection exists between them and the things they signify. Yet the belief that there is a kind of inherent fitness between the word and the thing it represents has long been curiously widespread. One of Plato's dialogues, *Cratylus,* explores this question. A speaker remarks, "I think the truest account of these matters is this, Socrates, that some power greater than human laid down the first names for things, so that they must inevitably be the right ones" (*Cratylus* 438).

The same belief lies at the root of primitive people's reluctance to tell strangers their names lest knowing should give others power over them. This tendency to identify words with things is apparent in many of the magical practices of primitive tribes and the beliefs of some religions. Early humans used verbal magical formulas in almost every aspect of life: to ensure success in the hunt, to produce rain, to guard against harm, to cast out evil spells, to put a curse on an enemy.

In religious rituals the repetition of certain words is frequently believed to cure bodily and spiritual maladies, ward off evil, and assure salvation. We also find in some religions the belief that certain words are so holy and sacred that they must never be spoken. Among the ancient Hebrews (as indeed still among Orthodox Jews today) the name of God, *Yahweh* or *Jehovah,* was "unnamable" and its utterance was forbidden in both speech and prayer. God could be referred to only through a circumlocution— *Adonai,* meaning "The Lord." Greek mythology demanded that Hades, god of the world below, be called Pluto ("the giver of wealth") in ordinary conversation, to avoid pronouncing the dread name *Hades*. The commandment that one should not "take the name of the Lord in vain" is probably the source of U.S. laws against blasphemy, which is still a crime in many places.

The matter of names arises in the Old Testament, where we read:

> And out of the ground the Lord God formed every beast of the field, and every fowl of the air. And brought them unto Adam to see what he would call them, and whatsoever Adam called every living creature, that was the name thereof. And Adam gave names to all cattle and to the fowl of the air, and to every beast of the field. (Genesis 2:19–20)

Mark Twain added his own sequel to this biblical story, in which he exploited the belief that a name has an inherent relation to the thing it names. When Adam couldn't think of a name for one of the animals, according to Twain, he appealed to Eve for help. "What name shall I give to this animal?" "Call it a horse," answered Eve. "But why a horse?" "Well," said Eve, "it looks like a horse, doesn't it?"

The error of identifying names with things has been the source of much humor. Countless parents have been asked by wide-eyed youngsters, "When I was born, how did you know I was Charlie and not some other little boy?"

It was an attempt to correct this tendency to assume that names "belong" by nature to the things they signify that led the British rationalist Thomas Hobbes to warn, in a now famous aphorism, "Words are wise men's counters, they do but reckon by them; but they are the money of fools" (*Leviathan,* pt. I, ch. 4). Abraham Lincoln tried to make the same point when he asked an audience, "If I call the tail of a horse a leg, how many legs will the horse then have?" "Five," they responded. "No," answered Lincoln. "Calling a horse's tail a leg doesn't make it one."

We smile at such stories, but the names we use to identify objects become, after long use, so associated in our minds with their objects that it is sometimes easy to forget for the moment that, for example, a pig is not so named because it is such a dirty animal. Or we say, "Speak of the Devil and he will appear"; we address the dice with endearing expressions such as "Little Joe" in order to influence them. In such cases, the false assumption that the symbol equals the thing leads us into a second false assumption—that the language we use somehow affects the thing we are referring to.

The false identity of names and things can affect people's actions as well as their thoughts. A noted historical example is the riots set off by introduction of the Gregorian calendar in 1582. After it had been agreed to initiate the new calendar by labeling October 5th (Old Style) as October 15th (New Style), angry mobs paraded the streets shouting, "Give us back our ten days!"

The eminent general semanticist S. I. Hayakawa, in a section of his book *Language in Thought and Action* appropriately entitled "The Word Is Not the Thing," cites other actions stemming from the false assumption that the symbol equals the thing:

> One is reminded of the actor, playing the role of a villain in a traveling theatrical troupe, who, at a particularly tense moment in the play, was shot by an excited cow-puncher in the audience. But this kind of confusion does not seem to be confined to unsophisticated theatregoers. In recent times, Paul Muni, after playing the part of Clarence Darrow in *Inherit the Wind,* was invited to address the American Bar Association; Ralph Bellamy, after playing the role of Franklin D. Roosevelt in *Sunrise at Campobello,* was invited by several colleges to speak on Roosevelt. (Fourth ed. New York: Harcourt Brace Jovanovich, 1978, pp. 25–26)

We are sufficiently prone to the error illustrated above to lead the Federal Trade Commission toward increasing regulation of the way in which manufacturers represent products. FTC regulations prohibit passing off domestic products as imported (as when rugs are labeled with foreign-sounding names like "Khandah" or "Calcutta," to give the impression that they are Oriental rugs when they are in fact domestic). Aspirin provides another example of the power of labels. In some states, aspirin is legally classified as a "drug" and can therefore be sold only by licensed pharmacists. In other states it is not classified as a drug and is more widely available. Yet residents of states where aspirin is now a drug might well decide one day to reclassify it (rename it) as "not a drug"—not because of anything new they have discovered about aspirin but simply because they wish to have it more widely available. Here again, language would be accommodating itself not to nature but to us and our special needs.

The art of renaming in order to hide unpleasant connotation produces *euphemisms.* A euphemism* is any agreeable expression that we substitute for one we find offensive. When we wish to avoid mentioning death, for example, we resort to euphemisms

---

Euphemism derives from the Greek *euphemos,* meaning "something that sounds good." (The Greek root *pheme* means "speech." *Eu* is "pleasing," as in *euphonious.*)

---

like *passed on, departed this life,* or *went to her reward,* trying thereby to transform the event, perhaps even to deny it. We call *third class* today *tourist class. A traveling salesperson* is now a *field representative,* a *janitor* is a *custodian,* and *garbage collectors* have become *sanitation engineers.*

Some subjects seem to provoke more euphemisms than others. Words associated with bodily functions are a case in point, as can be seen by reviewing the variety of terms which have been used for the *gong.* That is what the Anglo-Saxons appropriately called it, for it is a place you *go to.* But *gong* began to seem crude and was replaced with the Latin *necessarium.* In Shakespeare's day, if one wanted to be vulgar he called the place a *jakes,* that is, a *jacks,* which is the literal equivalent of the modern *john* or *johnny.* Later the room came to be called a *water closet* (just as we now call it a *bathroom*), which name was soon concealed in the initials

W.C. Those of more delicate taste in the meantime called it a *toilette.* When that term was shortened to the somewhat less delicate *toilet,* the place became a *powder room,* or a *men's room* or *women's room.*

In logic, as in all use of language, we need to remember that the use of a word, whether the word is "good" or "bad," cannot of itself guarantee the existence of characteristics implied by it. We shall have occasion to look more closely at euphemisms and their opposites when we examine logical errors in part two.

EXERCISES

The following advertisements try to exploit certain of our feelings by calling up associations in our minds. Analyze in each case what the association is and why it is (or is not) effective as a persuader. The first one has been analyzed.

9. To the 56,000,000 people who smoke cigarettes.
    Analysis: By calling up huge numbers of people, this headline in a cigarette ad is designed to ease fears that we as smokers, or potential smokers, may have about smoking. There are, after all, quite a number of people who still smoke—the implication being that it probably isn't as dangerous as has been claimed.

10. A great American soup. About as close as you can get to homemade without making it yourself.

11. One American custom that has never changed: a friendly social drink.

12. A pen can't tell you everything about a man. But it can tell a lot. The Parker 75 Classic Ball Pen.

13. The first completely new way to tell time in 500 years . . . invented and made in U.S.A.

14. The kind of car everyone is trying to build.

15. Army. Be all you can be.

16. Your family can be a CBS Software Family, too. All you need to spend more time together—and have more fun when you do—is CBS Software. No other software company can turn your family's computer into a center of family fun.

17. You know what they're saying. The new Sears is sassy. Ditto the new Sears Catalog. Proof, here we are. Cheek to cheek with le beau monde. The new Sears Catalog has

looks that define instead of follow. Brand names of the sort that sprinkle conversations. And enough ideas to be an awfully good read. So dismiss those quaint notions of the old general store in print.

18. Friends are worth Smirnoff. When the friends are close and the mood is right, the party starts in the kitchen. And, of course, Smirnoff Vodka is there. Because nothing but Smirnoff makes drinks that are as light and friendly as the conversation. Crisp, clean, incomparable Smirnoff.

19. The 1985 Continental. If it didn't have all the technology it has, it wouldn't have all the luxury you want.

20. 1985 Ninety-Eight Regency. The remarkable marriage of luxury and precise road management.

From memory, list ten euphemisms. Write a brief explanation of the first five on your list.

4. THE USES OF LANGUAGE

A fundamental use to which language can be put is to give information about the world: "That table is made of oak," for example, or, "I've lost my watch." Language whose primary intention is to transmit information tends to be descriptive, straightforward, and nonevaluative. But we may also use language to express our feelings or to evoke feelings in others. We exclaim, "What a lovely day!" not to inform others about the weather but to express our joy in it. Or we may say to our host, "It seems a bit chilly in here," not as a simple factual statement but in order to induce him or her to turn up the thermostat.

In short, language can be used not only *informatively* but also *ceremonially, emotively,* and *directively,* among a great variety of other uses. Each use has its own legitimate domain. What can be objectionable, however, is confusing one use with another. Some confusions of language and purpose are merely amusing. The salutation "How are you?" is not really a request for information about the other person's state of health. The proper answer is simply, "Fine, thank you," even if you are far from being so.

More serious than this confusion of the *ceremonial* use of language with the informative use is the confusion of the emotive with the informative. In such cases, the language used may be inappropriate to the purpose and may try to carry us to the goal by a circuitous route or to divert us from the goal entirely. William Jennings Bryan's famous "Cross of Gold" speech in opposition to the gold standard provides a vivid illustration of emotive language. Its rousing, emotionally charged last words were: "You shall not press down upon the brow of labor this crown of thorns. You shall not crucify mankind on a cross of gold!" Bryan's appeal traded on deep feelings associated with the religious symbolism of the crucifixion, the cross, and the crown of thorns. By associating the issue of the gold standard with these religious symbols, the speech attempted to transfer to the gold symbol the powerful emotions with which these religious symbols had long been invested. The key element to isolate when ascertaining whether a given use of language is appropriate is its intention. Is the primary purpose of the communication to inform? Is it to express certain emotions? Is it designed to move us to do something? Is it perhaps all of these at once? Having established intention, consider whether the language is in keeping with that intention. If facts alone are at stake, an overdose of emotionally tinged language can only be a hindrance and should be regarded with suspicion. Even eloquence should be shunned where information is the sole purpose. For facts can best be dealt with if stated in a plain and objective manner.

The fact that emotional language is usually to be avoided in logic does not mean that it is in any way an inferior kind of language. A poet, whose primary domain is not reason, is more likely to achieve his or her purpose by choosing words that we can appreciate with our feelings rather than our reason. This may be the meaning of T. S. Eliot's observation that poetry can communicate before it is understood. It is probably also responsible for the fact that attempts to reduce a poem to terms strictly understandable by our reason tend to destroy qualities it had as a poem. It is said that Alfred Noyes Tennyson objected to William Wordsworth's line, "And sitting on the grass partook the fragrant beverage drawn from China's herb," asking, "Why could he not have said, 'And sitting on the grass had tea'?" Tennyson's proposal, however, robs Wordsworth's line of its poetry, though it manages to retain the semantic content. The purpose of poetry, as of all uses of language, is to employ words that are appropriate for the task at hand.

The emotive value of a speech or essay is, however, irrelevant to its logical value. When we are reading or listening critically from a logical point of view, we are interested only in the informative content of statements and the support some statements give to others. It is, to be sure, a sign of sophistication to be able to recognize the different uses of language and to judge instances of each by the criteria appropriate to them. But, while granting that no speech or essay is worthless merely because it is emotive, we need to keep in mind that the emotive force of an argument has no bearing whatsoever on its validity.

EXERCISES

In the following items, identify the chief purpose of the discourse. Is it to inform? To express feelings? To direct? To engage in a certain ceremony? If there is more than one purpose, which is dominant? Finally, analyze whether the language used in each case is appropriate for the purpose. (See section 4.)

* 21. It's a pleasure to meet you.

* 22. Put your hands up!

23. I invite you to sit down in front of your television set when your station goes on the air and stay there without a book, magazine, newspaper, profit-and-loss sheet or rating book to distract you and keep your eyes glued to that set until the station signs off. I can assure you that you will observe a vast wasteland. (FCC Chairman Newton Minow)

24. I've cried my poor eyes out over your horrid play, your *heavenly* play. My dear, and now! How can I go out to dinner tonight? I must keep my blue glasses on all the while for my eyes are puffed up and burning. But I can scarce keep from reading it all over again. Henry would not care for that play, I think. I know he would laugh. And that sort of thing makes me hate him sometimes. (From a letter written by Ellen Terry to George Bernard Shaw, October 19, 1896, after she had read Shaw's *Candida*)

* 25. Shut the window.

26. As far as the laws of mathematics refer to reality, they are not certain; and as far as they are certain, they do not refer to reality. (Albert Einstein)

27. Your party was out of this world!

28. Make sure your names are on your booklets before you hand them in.

29. The heavens declare the glory of God; and the firmament sheweth his handiwork. Day unto day uttereth speech, and night unto night sheweth knowledge. There is no speech nor language, where their voice is not heard. (Nineteenth Psalm)

30. For it is one thing to employ a concept efficiently, it is quite another to describe that employment; just as it is one thing to make proper use of coins and notes in marketing and quite another thing to talk coherent accountancy or economics. Efficiency at the one task is compatible with incompetence at the other, and a person who is not easily cheated when making purchases or getting small change, may easily be taken in by the wildest theories of exchange-values. (Gilbert Ryle, *Dilemmas*)

31. The human mind has no more power of inventing a new value than of imagining a new primary colour, or, indeed, of creating a new sun and a new sky for it to move in. (C. S. Lewis)

32. O western wind, when wilt thou blow
That the small rain down can rain?
Christ, that my love were in my arms
And I in my bed again!

(Anonymous medieval lyric)

33. When a true genius appears in the world, you may know him by this sign, that the dunces are all in confederacy against him. (Jonathan Swift)

34. If you want to print with the NEC letter quality printer, you MUST use the "Wordstar For NEC Spinwriter" disk.

35. If there is anything that marks a town as being a genuine hicksville, it is the innocent belief that a new domed stadium is the height of progress. (Mike Royko)

36. I hope in these days we have heard the last of conformity and consistency. Let the words be gazetted and ridiculous henceforward. Instead of the gong for dinner, let us hear a whistle from the Spartan fife. Let us never bow and apologize more. A great man is coming to eat at my house. I do not wish to please him; I wish that he should please me. (Ralph Waldo Emerson)

5. AMBIGUITY AND VAGUENESS

Ambiguity and vagueness are alike in that they are both characteristic of imprecise language. The distinction between them is instructive, however. A word or expression is *ambiguous* if it has more than one distinct meaning. It is *vague* if its meaning is unclear. Ambiguous language confronts us with several meanings, of which the correct one is difficult to determine, while vagueness confronts us with the task of hunting for any meaning at all. To say, "That was the most shocking book I ever read" would be an ambiguous statement. "What a book!" would be vague.

While the meaning of vague words is doubtful, that of ambiguous words is double (or even triple and beyond). In many cases the context in which the ambiguous language occurs will determine which of its meanings is intended. This importance of context is not necessarily a defect of language but rather an indication of its flexibility. Until put to work in a specific context, some perfectly serviceable words necessarily remain indeterminate. If they remain indeterminate once placed in a context, however, such words must be considered ambiguous.

The case with vagueness is different. Some words are always vague, regardless of context, for their meanings are not merely indeterminate but indefinite. The word *rich,* for example, is always vague because it is never clear how much wealth a person must have before it is appropriate to call him or her rich; the word *cleave,* on the other hand, is potentially ambiguous because, depending upon the context, it may mean either "bring together" or "tear apart."

Unlike ambiguity, which has only one cause (the fact that context alters the meanings of words), vagueness has several causes. Sometimes vagueness simply reflects our own confusion. At what age, for example, can a person properly be said to be "middle-aged"? And how bald does a person have to be before we may correctly call that person "bald"?

At other times, however, vagueness is due not to a lack of clarity in us but rather to the shopworn nature of language itself. Certain words have become blunted from overuse, losing the precision they once had. Words that fall into this category include *fine, inspiring, great, elegant,* and the like. In addition, some vague words have acquired so many meanings that they have lost whatever precision they once had. Examples are *democracy, communism, art, progress, culture,* and even the word *vague* itself. In such cases the terms need to be defined before they are used. Finally, there are some words which are both ambiguous *and* vague—the word *artist,* for example.

EXERCISE

Make up a list of vague words, indicating those which are vague because (1) we ourselves are unclear about the things involved (e.g., *middle-age*); because (2) they have been overused in our language (e.g., *elegant*); or because (3) they have come to stand for too many things (e.g., *democracy*). (See section 5.)

6. VERBAL DISPUTES

In our exploration of the role of language in argument, it is important to make a distinction between two types of disputes, *real disputes* and *verbal disputes,* before moving on to analyses of specific disputes in part two. A dispute is said to be real when one party believes that a certain statement is true while another party believes the statement is false. Real disputes arise when genuine differences of opinion exist regarding matters of fact.

Verbal or linguistic disputes, on the other hand, occur when one party believes that a certain statement is true while another party believes that *another* statement is false. Rather than a difference of opinion over a single statement, there is a different view of what is at issue. In such a case the parties argue at cross-purposes because neither party realizes that the argument is not over the same statement.

Because verbal disputes often look like ordinary factual disputes, we are often mistakenly tempted to settle them in the manner in which we would settle a real dispute. Of course, our attempts prove futile, for facts cannot settle verbal disputes. They can be settled only by recognizing that they are simply verbal and by untangling the verbal knots in which both parties are caught up.

In verbal disputes the parties may or may not be in agreement regarding the facts of the case, but, because of the manner in which each of them understands key terms in the dispute, they cannot tell what their dispute is really about—let alone whether they agree or disagree about it. Arguments about "God" are frequently of this nature. In such arguments it is rare that a disputant comes to see that, since the meanings each party has attached to the term *God* are different, they have not been arguing about the same thing at all. One theological student is reported to have burst out after a typical argument of this sort: "Now I see! Your God is my Devil, and my Devil is your God!"

We have seen throughout this chapter that words do not serve merely to convey information but also to convey feelings and attitudes. Since the same words that convey information are sometimes used to express feelings and attitudes, occasions for verbal misunderstanding are frequent. In order for there to be a true meeting of minds, the parties to the dispute must agree regarding not only the descriptive meanings of their words but their expressive meanings as well.

A case in point is the word *aggression,* particularly as it is used in international disputes. While both parties or nations may appear to agree as to the expressive meaning of the word—it is a "bad" word for both—they do not agree regarding its descriptive meaning. Thus for one party, *aggression* does not seem to include propaganda, sending military equipment or intelligence agents or reports into another country, providing foreign armies with a body of instructing officers, and so on. For an opposing party, however, all such acts might properly be called acts of aggression. Yet, to accuse one party of hypocrisy when it denied being guilty of "aggression" is to be blind to the underlying semantic problems that lie at the root of this linguistic misunderstanding.

One can imagine another type of verbal dispute over aggression, in which the parties agree upon the meaning of the term but deny that it represents anything other than enlightened self-interest rather than something a nation ought to be ashamed of. Verbal disputes, then, can arise in either of two ways: (1) when both parties, although sharing the same emotional dimension of the issue, differ in their understanding of its descriptive dimension; or (2) when both parties, although in agreement on the descriptive dimensions of the issue, have entirely different responses regarding its emotional dimension. Both types of verbal disputes are often accompanied by a sense of puzzlement. The parties feel unsure of where they are in the argument or how they should proceed. The best procedure in such cases is to face one's bafflement and try to get one's bearings by inquiring of the other party just what it is he or she means by the term at issue.

A dispute may be verbal in still a further sense, as in the following question familiar to beginning students of philosophy: Does the fall of a tree in an uninhabited forest cause a sound? There is clearly only one key term in this question and that is the term *sound.* Since the answer we will be inclined to give to this question depends on the meaning we attach to that term, it is important that we establish at the outset just what that meaning is. Unfortunately, as used in this context, the term is ambiguous, because it may mean either (a) an audible sensation or (b) a series of waves in the air capable of causing an audible sensation when they impinge on the human ear. The question, therefore, of whether a tree falling in an uninhabited forest makes a sound can carry either a "yes" or a "no" answer—either of which is equally justifiable depending on the meaning we attach to the key term. "Yes, it makes a sound," we might answer, "if by sound we mean *'air vibrations.'* Such vibrations obviously occur whether or not anyone is there to receive them." We might, on the other hand, reply, "No, it makes no sound, if by sound we mean the *experience* of certain sensations. Since no one is there to experience such sensations, no sound is made." Such a dispute is thus essentially verbal, for the argument produces the illusion of there being a difference on one issue between the parties when in fact they are arguing separate issues.

EXERCISES

Analyze the way in which the issues are or can be resolved in the following passages. (See section 6.)

37. We have been told by popular scientists that the floor on which we stand is not solid, as it appears to common sense, as it has been discovered that the wood consists of particles filling space so thinly that it can almost be called empty. This is liable to perplex us, for in a way of course we know that the floor is solid, or that, if it isn't solid, this may be due to the wood being rotten but not to its being composed of electrons. To say, on this latter ground, that the floor is not

solid is to misuse language. For even if the particles were as big as grains of sand, and as close together as these are in a sandheap, the floor would not be solid if it were composed of them in the sense in which a sandheap is composed of grains. Our perplexity was based on a misunderstanding: the picture of the thinly filled space had been wrongly *applied.* For this picture of the structure of matter was meant to explain the very phenomenon of solidity. (Ludwig Wittgenstein, *The Blue and Brown Books*)

38. Jesus said: "Truly, truly, I say to you, unless one is born again, he cannot see the kingdom of God." Nicodemus said to Him, "How can a man be born when he is old? He cannot enter a second time into his mother's womb and be born, can he?" (*John* 3:3)

7. DEFINITION

Anyone who presents an argument for serious consideration is under an obligation to state his or her conclusion and premises clearly. One way to make an argument clear is to give *definitions* of key terms. A definition is a statement that one word or phrase has the same meaning as another word or phrase. It consists of three elements: (1) the defined expression, which logicians call the *definiendum;* (2) the defining expression, which logicians call the *definiens;* and (3) an assertion or stipulation that the defined expression has the same meaning as the defining expression. For example, if we accept the authority of *Webster's New Collegiate Dictionary,* the word *extensor* may be formulated thus: *Extensor* means "a muscle serving to extend a bodily part (as a limb)." The defined expression (definiendum), a word which is unfamiliar to most of us, is the term *extensor.* The defining expression (definiens), a phrase which consists of words we are familiar with, is "a muscle serving to extend a bodily part (as a limb)." The verb *means,* which connects the defined expression and the defining expression of this definition, may be paraphrased "has the same meaning as."

Logicians refer to definitions like this one from *Webster's New Collegiate Dictionary* as *reportive definitions* and contrast them with *stipulative definitions.* A reportive definition is a statement that a word or phrase is used in a certain way by a specific linguistic group, for example, speakers of standard English. A stipulative definition, on the other hand, is a statement by a speaker or writer that he or she intends to use a word in a certain way. Reportive definitions can be judged as either true or false. Stipulative definitions can be judged only as either helpful or unhelpful. Often the most helpful kind of definition for the purposes of argument is a hybrid of the reportive and the stipulative. Let us call it the *explicative definition.*

A word that is subject to an explicative definition has a core of agreed upon meaning but a penumbra of vagueness that must be clarified if the term is to be useful in argument. An explicative definition accepts the agreed upon usage of its defined expression but goes on to stipulate how that expression is to be used in cases not made clear by dictionary reports of its agreed upon usage. Suppose, for example, a person were arguing that the Common Market nations should form a federal union. This person would have to define what he or she means by *federal union.* The definition would have to correspond to the way the expression is used by speakers of English. For example, the person could not leave out of the defining expression the idea of independent political entities giving up some of their sovereignty to a central government. A mere report of the standard usage of *federal union,* however, would not make clear how much and what kind of sovereignty a group of independent political entities would have to give up to a central government in order to count as a federal union. On this point, the person would have to stipulate how the term will be used in his or her particular argument.

Webster's definition of *extensor* is an example of what logicians call definition *per genus et differentia.* A definition *per genus et differentia* is one whose defining expression refers the item to its generic class and then distinguishes it from all other kinds of items in that class. For example, Webster classifies an extensor as a muscle and then goes on to indicate the specific kind of muscle it is. Not all words can be defined in this way, but many can, and a good definition *per genus et differentia* is a precise tool of thought.

The marks of a good definition should be obvious. The defining expression must be clearer to the audience or reader than the defined expression. It is no use defining a key term of an argument in words that are themselves vague or obscure, for the point of giving definitions is to eliminate vagueness and obscurity. In addition, the defining expression should be stated in terms that are grammatically parallel with the word or phrase defined, for, technically, the defining expression of any definition should be substitutable in any linguistic context for the defined expression. The above definition of *extensor* satisfies this requirement, since *extensor* is a noun and "a muscle serving to extend . . ." is a noun phrase.

A good reportive definition must be *accurate* as well. A definition is accurate if and only if its defining expression applies to all and only the items to which its defined expression applies. If the defining expression of a definition applies only to some of the items correctly named by the defined expression, then we say that the definition is too *narrow.* Suppose, for example, that someone were to define the term *education* by saying that it means "the development of the ability to think clearly." The defining phrase of this definition does cover part of what the defined expression correctly refers to. But because it does not apply to other clear instances of education—for example, the development of literary taste or historical understanding—the definition is too *narrow.* Suppose, on the other hand, someone were to define *education* by saying that it means "the total adjustment of the individual to his or her total environment." This definition is too *broad.* Its defining expression applies to activities and processes that we do not count as referents of *education,* for example, the reflex of blinking.

The way to test the accuracy of a definition, then, is simple: it must be put to the test of instances. One should try to think of something to which the defined expression refers that the defining expression does not cover, or vice versa. If that can be done, then the definition is inaccurate and therefore, for the purpose of precise argument, worthless.

In writing an argumentative essay, we normally need to devote more than a single sentence to the definition of each key term. It often requires a paragraph—sometimes an entire section of an essay—to develop a definition. The following three paragraphs from Cleanth Brooks and Robert Penn Warren's *The Scope of Fiction* are merely the starting point of their attempt to clarify what we mean when we refer to the *plot* of a piece of fiction:

> Let us now try to make a more systematic statement about the nature of the plot. We may begin with the most off-hand notion: Plot may be said to be what happens in a story. It is the string of events *thought of* as different from the persons

involved in the events and different from the meaning of the events. We make such a distinction even though we know that, in fact, we cannot very well separate an act from the person who commits it, or from its meaning as an act. The distinction is one we make in our heads and do not find ready-made for us in fiction. In order better to understand the nature of a story itself, we analyze the unity which is the story and which is what we actually experience before the process of analysis begins.

Plot is what happens in a story—that is a good rough-and-ready way to put the matter. But let us go a step farther. Plot, we may say, is the structure of an action as presented in a piece of fiction. It is not, we shall note, the structure of an action as we happen to find it out in the world, but the structure within a story. It is, in other words, what the teller of the story has done to the action in order to present it to us. Let us hang on to this distinction between a "raw" action—action as it occurs out in the world—and an action in a story, that is, action manipulated by the teller of the story.

Here we are using the word *action* . . . in a special way. We do not mean a single event—John struck Jim with a stone, Mary put the book on the shelf. We mean a series of events, a movement through time, exhibiting unity and significance. It is a series of connected events moving through three logical stages—the beginning, the middle, and the end. (Englewood Cliffs, N.J.: Prentice-Hall, 1960, pp. 51–52)

Whether we describe what Brooks and Warren are doing as giving a definition of the literary term *plot* or as clarifying the concept of plot in fiction, their discussion is a clear example of what one very often must do in order to eliminate from speech or writing, and therefore from thinking, any ambiguity and vagueness.

EXERCISES

Define the following terms: *bachelor, niece, triangle, democracy, freedom, education, morality, religion, love, friendship, truth, beauty, wimp.*

Example: Democracy means "a form of government in which political decisions are made by the vote of the citizens or their representatives."

8. THE ART OF PLAIN TALK

Clarity in the presentation of an argument cannot be achieved merely through the use of definitions. It also requires *style*. Good style is a matter of saying clearly, simply, appropriately, and concisely what it is we wish to say. If a thing can be put simply, why put it otherwise? Even if a subject is complex, we should, at least, strive to express it clearly—with as much precision as we can command. To achieve good style is to exercise restraint; one must avoid using many words where fewer will do, big words where smaller ones will do. One must curb the desire for elegance or picturesqueness and cultivate the desire to speak plainly.

In Shakespeare's play *As You Like It,* the character Touchstone provides an illustration of what plain style is *not:*

> Therefore, you clown, abandon—which is in the vulgar, leave—the society—which in the boorish is company, of this female—which in the common is woman; which together is, abandon the society of this female, or, clown, thou perishest; or, to thy better understanding, diest; or, to wit, I kill thee, make thee away, translate thy life into death. (act 5, sc. 1)

The chief virtue of any language is that it enables its users to say exactly what is meant, no more and no less.

One impediment to good style is the *cliché.* A language tends to accumulate these in abundance over its long history, but in this case we would do well not merely to use them with caution but to renounce them entirely. A cliché is a trite expression conveying some "truth" that is so overused that it has become practically meaningless. If we were to argue, "Haste makes waste, so let us move slowly in righting this social injustice," our use of a cliché as the premise would weaken the argument. Because clichés are short, often witty, and seemingly apt, they are always dangerous, for they encourage thoughtlessness by obscuring the fine points of an issue.

The best defense against cliché-riddled thinking is to remember that, for every such popular maxim, another can be quoted which directly contradicts it.

"Sure — but can you make him drink?"

"The Far Side" cartoon by Gary Larson is reprinted courtesy of Chronicle Features, San Francisco.

Testing a cliché.

a) Two heads are better than one.
b) Two captains sink a ship.

c) Don't change horses in midstream.
d) It's time for a change.

e) Better to be safe than sorry.
f) Nothing ventured, nothing gained.

A cliché can probably be found to prove—or disprove—almost anything.

The most serious impediment to good style, however, is *jargon,* aptly characterized by Professor Donald Hall as "the clichés that belong to a particular profession." Take, for example, the following sentence from a publication of the National Science Foundation:

> The award of which this attachment is an integral part constitutes acceptance by the National Science Foundation (NSF) of the proposal referenced in the award letter and its agreement to assist in the financial support of the project described in that proposal at the level of effort stated in the award letter which reflects such revised proposal budget as may have been submitted.

What makes this sentence objectionable is not the use of technical terms which are unfamiliar to the layperson, for it contains no such terms, and even if it did, that fact would not in itself be grounds for criticizing it. Rather, the sentence offends because of its manner of expression, which both obscures and inflates the meaning of a simple thought. Note the pretentious diction ("an integral part," "constitutes," "referenced"), the excessive use of prepositions, and the absence of a straightforward pattern of subject, active verb, and object. These stylistic features of this sentence are what is meant by jargon. Jargon is characteristic of the bureaucratic mentality, but it is also pervasive in the writing of journalists, scholars, and students. Any writer who presents an argument for serious consideration is under an obligation to avoid jargon.

EXERCISE

> How many clichés can you remember, by trying to think of one starting with each letter of the alphabet? Continue the list as started below. (See section 8.)
>
> Absence makes the heart grow fonder.
> Birds of a feather flock together.
> Charity begins at home.
> Don't count your chickens before they hatch.

SUMMARY

This chapter on language has formed a bridge between logic as the study of argument, which was the subject of chapter 1, and the analyses of specific arguments, which will follow in part two. We saw that the language employed in any statement determines the meaning of that statement. Language was seen as intimately related to thought, even as a shaper of thought.

Language was seen to be symbolic as well, in that all words are conventional signs for the things they signify—as opposed to natural signs, which are parts or symptoms of what they signify. The tendency to equate words with the things they symbolize was shown to foster confusions of meaning and to give rise to euphemisms, in which things are given new names in order to disguise negative features.

Other confusions of meaning were found to result from a mismatch of type of language—such as informative, emotive, directive, or ceremonial—with the intention of the communication. Two defects which mar informative discourse were described— namely, ambiguity, where more than one meaning may attach to particular words or phrases; and vagueness, where the meaning is simply unclear. We noted that disputes can be either real, in which case there is genuine disagreement over the same issue, or verbal. In verbal or linguistic disputes the parties differ on which question is being argued, usually without realizing it. They may differ on the descriptive meaning of the language used or on its emotional meaning.

The use of definitions was recommended as one way of avoiding vagueness in an argument, not to mention ambiguity and verbal disputes. It was pointed out, however, that the use of definitions is not sufficient to achieve clarity. A plain style is required when our intention is to convince others that our opinions are rational and worthy of acceptance. Such a style, which avoids clichés and jargon, says simply and directly just what we want to say.

ANSWERS TO STARRED EXERCISES

1. This statement raises the issue of whether or not thought is independent of language. If it is, then the same unit of thought should be expressible in different languages. If, however, the expression of a unit of thought is inextricably connected with its meaning, then no two languages can say exactly the same thing, since no two languages have exactly the same forms of expression.

4. Alice has drawn attention to two facts about language that follow from the fact that language is conventional. Because language is conventional, the words we use have the meanings they have because we have given them those meanings. But at the same time, again because language is conventional, no one person can make words mean anything he or she chooses, for, in order to communicate in language, that individual must use words in the same way others do.

21. Ceremonial. One can and normally does utter this expression without intending to express or evoke any sort of feeling whatsoever.

22. Emotive and directive. The end punctuation justifies classifying this sentence as emotive. The mode of the sentence, the fact that it is an imperative, justifies classifying it as directive.

25. Directive. Compare with question 22. It is imperative but not exclamatory.

SUGGESTED READINGS FOR PART ONE

Most of these works on logic and on language are relatively easy to read and have become widely popular.

Max Black, ed. *The Importance of Language.* Ithaca, N.Y.: Cornell University Press, 1969. Selected papers on the philosophy of language, edited by a noted American analytic philosopher.

Lewis Carroll. *Alice's Adventures in Wonderland.* New York: New American Library, 1960. Readers can learn a great deal about the nature of language if they can figure out precisely why Alice and the creatures of Wonderland continually fail to communicate.

Edward T. Hall. *The Silent Language.* New York: Doubleday Anchor Book, 1973. An account by an American anthropologist and linguist of how we "talk" to each other without the use of words.

S. I. Hayakawa. *Language in Thought and Action.* 4th ed. New York: Harcourt Brace Jovanovich, 1978. One of the foundational works in general semantics (concerning the way language orders our lives), written by one of the main architects of that discipline.

Richard Lanham. *Revising Prose.* 2nd ed. New York: Macmillan, 1987. An excellent practical guide to writing clearly. Contains a formula for reducing the "lard factor" in one's prose style.

Richard Mitchell. *Less Than Words Can Say.* Boston: Little, Brown, 1979. A wise and humorous analysis of the relationship between linguistic and intellectual slovenliness.

George Orwell. *Nineteen Eighty-four.* New York: New American Library, 1949. In this dystopia, Orwell shows, among other things, how by controlling the language of its citizens a totalitarian government succeeds in controlling their thoughts and therefore their behavior.

George Orwell. "Politics and the English Language," in *A Collection of Essays by George Orwell.* New York: Doubleday & Co., 1953. Traces the decline in standards of speech and writing evident in our culture to the pernicious influence of a debased politics. Contends that the reform of our speech habits is a political act.

John Simon. *Paradigms Lost: Reflections on Literacy and Its Decline.* New York: Clarkson N. Potter, Inc., Publishers, 1980. An energetic critique of what the author, who is a major critic of the arts, perceives as the slovenly speech habits that pervade the public discourse of our time.

Frank Smith. *Essays into Literacy.* London: Heinemann Educational Books, 1983. These essays by a brilliant British educationalist on the acquisition of literacy explode prevailing myths about how children learn to read and write. The author explores the meaning of literacy in the life of the individual and exposes the unsoundness of current methods of teaching language in the schools.

Lev Semenovitch Vygotsky. *Thought and Language.* Edited and translated by Eugenia Hanfmann and Gertrude Vakar. Cambridge, Mass.: M.I.T. Press, 1962. A highly original discussion of the nature of language and how we come to acquire and master it. By a brilliant Russian linguist and psychologist who died of tuberculosis at the age of thirty-eight.

Benjamin Lee Whorf. *Language, Thought and Reality.* Edited by John B. Carroll. Cambridge, Mass.: M.I.T. Press, 1964. How language shapes our thoughts and perceptions. By one of the most articulate modern spokesmen for the theory of linguistic relativity.

Joseph M. Williams. *Style: Ten Lessons in Clarity and Grace.* 3rd ed. Glenview, IL. Scott, Foresman, 1989. A practical manual on writing clear, strong English prose. Each lesson is a specific practical hint on developing a style free from the sort of wordiness and awkwardness that obscures meaning and impedes clear thought.

John Wilson. *Thinking with Concepts.* Cambridge: Cambridge University Press, 1966. An excellent introduction to the practice of linguistic analysis, i.e., the method of clarifying the meaning of abstract words which signify ideas pervasive in our thought and conduct.

Part Two

Informal Fallacies

Part two examines common types of fallacious arguments. A *fallacy** is an argument that is unsound. Although logically

---

Fallacy probably derives from the Greek *phelos*, meaning "deceitful," which is thought to be related to our word *fail*.

---

flawed, fallacies are usually persuasive and often upon first examination appear sound in form and content.

The fallacies we will explore are called *informal fallacies* because their persuasiveness rests on "material" (as opposed to "formal") factors, such as ambiguity, the use of slanted language, and prejudice. The error in these arguments, in other words, lies in their content, not in their form or structure. We deal with informal fallacies by identifying and clarifying their vague or ambiguous statements, by making explicit their questionable assumptions, and by exposing their biases. It is unfortunately sad that the respects in which these arguments are unsound are frequently what commends them to their hearers.

Logicians of all periods have studied these fallacious arguments, which are often subtle and complex. The first to classify fallacious arguments was Aristotle, who founded the science of logic in the fourth century B.C. Aristotle divided the common fallacies into two groups: those having their source in language, which included the fallacies of ambiguity, and those having their source outside language, which accounted for all the other fallacies. Although many have tended to follow Aristotle's classification, the treatment of some fallacies has changed with time, and new fallacies have been identified. Others have argued that no satisfactory classification of fallacies is possible, since the ways to error are so numerous and complex. Still others have asserted that, as the study of correct reasoning, logic should not concern itself with faulty reasoning. But this argument is itself probably a fallacy, since familiarity with common logical errors helps us defend against fallacious thinking— both in others' arguments and in our own—and thus promotes the cause of correct reasoning.

In this book we shall depart from Aristotle's twofold classification by emphasizing how all fallacies have their source in some dimension of language. The organization of the treatment of fallacies in part two is thus threefold, corresponding to three ways in which the language used in a fallacious argument can be seen as the source of error. In chapter 3 the subject is fallacies of ambiguity, where it is the multiple meaning of the words employed that is the source of the fallacy. In chapter 4, on fallacies of presumption, the error stems from the similarity of the fallacious arguments to correct arguments—inviting us to "presume" that the reasoning must be correct because it contains language resembling that of correct arguments. The final group, in chapter 5, consists of fallacies of relevance, in which irrelevant language is introduced in order to buttress an emotional, rather than a logical, appeal. Thus, whereas Aristotle traced error in argument directly to language only in the case of fallacies of ambiguity, we shall find it useful to stress the linguistic nature of the error in fallacies of presumption and relevance as well—in fallacies of presumption because they ape the language of valid arguments and in fallacies of relevance because they employ emotional language.

It is not clear who first proposed and used this threefold division. That categorization nevertheless fits the subject very well. For logic is the study of argument, and before giving our assent to an argument, we should always make sure we are clear about the following three things:

1. Is what the argument asserts clear?
2. Are the facts in the argument correctly represented?
3. Is the reasoning in the argument valid?

The three categories of fallacies are tied to these three aspects of argument. The first set (the fallacies of ambiguity) deals with arguments that fail to meet the challenge of the first question (Is the argument *clear?*); the second (the fallacies of presumption) deals with arguments that fail to meet the challenge of the second question (Is what the argument asserts *true?*); and the third (the fallacies of relevance) deals with arguments that fail to meet the challenge of the third question (Is the argument *valid?*).

In the course of illustrating these three types of fallacies, it will sometimes be useful to examine somewhat absurd examples that no one is ever likely to commit. Such examples serve a purpose similar to that of the telescope or microscope in other fields: they magnify the nature of the subject under study so that we can see it more clearly. In other cases a humorous example may be used to illustrate what is in reality an extremely serious fallacy. Here again, such light examples help highlight the nature of the error, but they mustn't lead us to think that fallacies are not dangerous. Adolf Hitler misled an entire nation with fallacious propaganda. Liberally educated persons of all times and places need to appreciate the serious consequences which can result from fallacious reasoning.

"Hang him, you idiots! Hang him! . . . 'String-him-up' is a figure of speech!"

3

Fallacies of Ambiguity

Fallacies of ambiguity are arguments that are unsound because they contain words that, either singly or in combination, can be understood in more than one sense. We observed in chapter 2 that our language contains many ambiguous words and expressions, those having more than one meaning. When ambiguity is introduced in argument, it always weakens the argument. We shall consider in this chapter six fallacies: those of amphiboly, accent, hypostatization, equivocation, division, and composition. Although ambiguity gives rise to all six, the specific ambiguity in each case is different. In amphiboly, we shall see that it is ambiguity of sentence structure that gives rise to the fallacy. In the case of accent, the ambiguity lies in the stress or tone employed. In hypostatization, ambiguity results from the use of a word that can properly refer only to abstractions as if it could also refer to concrete entities. In equivocation, ambiguity stems from the fact that the words used have more than one correct sense and could have various meanings depending upon their context. In division and composition, the ambiguity lies in confusing the collective with the distributive senses of terms. Fallacies of ambiguity can be witty and charming in those instances where we can assure ourselves that they are not dangerous. In other instances they can leave profound questions unsettled.

Table of Fallacies of Ambiguity

Fallacy	Definition/Hints	Example/Method
Amphiboly	An ambiguity caused by faulty sentence structure (*Focus*: The fallacy involves the whole sentence and doesn't turn on one word.)	"With her enormous nose aimed toward the sky, my mother rushed to the plane." (Clarify the ambiguity: Whose nose? The mother's or the plane's?)
Accent	A statement that is ambiguous because (1) its intended tone of voice is uncertain; (2) its stress is unclear; or (3) it is torn from context	(1) *Sarcastic or serious?* "It is impossible to praise this book too highly." (2) *Which word?* "Jones thinks McIntosh will succeed." (3) *In the spirit of the original?* "Will Rogers never met George McGovern."
Hypostatization	The treatment of abstract terms like concrete ones, sometimes even the ascription of humanlike properties to them (Like personification)	"This system stifles creativity and suppresses those who work hardest." (Observe subject and verb.)
Equivocation	An ambiguity caused by a shift between two legitimate meanings of a term (*Focus*: The ambiguity turns on one word or short phrase. Contrast with AMPHIBOLY.)	"If you believe in the miracles of science, you should also believe in the miracles of the Bible."

Table of Fallacies of Ambiguity *(Continued)*

Fallacy	Definition/Hints	Example/Method
Division	The assumption that what is true of (1) the whole or (2) the group must be true of the parts or members (Trying to "divide" what is true of the whole among its parts)	"I can't break this bundle of sticks, therefore, I cannot break any one of them." "This is the richest sorority on campus; Mary, who is a member of it, must therefore be one of the richest young women on campus."
Composition	The assumption that what is true of (1) a part of a whole or (2) a member of a group must be true of the whole or the group (Trying to "compose" the whole out of its parts)	(1) "The piece of pie I've been served is wedge-shaped and so is my neighbor's. They must have come from a wedge-shaped pie." (2) "Some day man will disappear from earth, for we know that every man is mortal."

1. THE FALLACY OF AMPHIBOLY

The *fallacy of amphiboly** is the product of poor sentence structure. It results when words are incorrectly or loosely grouped in a sentence, giving rise to a meaning not intended by the author. The

---

*Amphiboly comes from the Greek *ampho*, meaning "double" or "on both sides." In origin, it is closely related to *ambiguity*.

---

following are examples of the incongruity that amphiboly sometimes entails.

a)   With her enormous nose aimed at the sky, my mother rushed toward the plane.
b)   Clean and decent dancing, every night except Sunday. (Roadhouse sign)

Simple errors like these are not likely to deceive anyone, yet they illustrate the confusion that careless sentence structure generates. Thus, in the case of argument *b* above, were one to believe that special performances could be expected on Sundays, one would be deceived into believing something as true which is not. It is in this sense that these sentences are fallacies.

Such potentially misleading sentences are often the product of improper punctuation, pronouns having ambiguous antecedents, or dangling modifiers.

c)   If you don't go to other people's funerals, they won't come to yours.

Sometimes the error results from using fewer words than are required to establish the context.

d)   We Dispense with Accuracy. (Druggist's sign)
e)   Just received! A new stock of sports shirts for men with 15 to 19 necks. (Advertisement)

The classified advertisement columns of daily newspapers are a rich source of such careless writing, as in the case of the fellow

Illustration by Ben Santora

Words grouped together in ways that create ambiguous meaning contribute to the fallacy of amphiboly.

who advertised his dog for sale by saying it "eats anything and is very fond of children."

Because they must compress as much meaning as possible into few words, newspaper headlines are prone to errors of ambiguity.

f) Killer Says Dead Man Was Chasing Him with Drawn Razor
g) Dock Workers Set to Walk Out in Atlantic Ports

Posters, signs, and hastily written announcements, all of which use compressed language, as newspaper headlines do, are other sources of frequent amphiboly.

A few moments of thought could have avoided these errors, the cause of which is not lack of space but simple carelessness.

h) The marriage of Miss Anna Black and Mr. Willis Dash, which was announced in this paper a few weeks ago, was a mistake and we wish to correct it.
i) Police authorities are finding the solution of murders more and more difficult because the victims are unwilling to cooperate with the police.

The assumption that readers "will know what I mean," even if I don't express it clearly, is more dangerous the more complex my subject is.

Although we generally understand by amphiboly a fallacy aris-
ing from faulty construction of a single sentence, it can also result
from the incongruous juxtaposition of two sentences.

j)   My husband took out an accident policy with your company,
      and in less than a month he was accidentally drowned. I
      consider it a good investment. (Testimonial for an insurance
      firm)

Even if we grant that the incongruity of amphibolous fallacies is
unintentional, our respect for the persons committing the fallacies is undercut. For if the authors had been careful enough to
anticipate that their statements were open to humorous interpre-
tations, they would certainly have revised them.

Belonging to a rather different category is the conscious and
intentional use of the device. An example is the Witch's prophecy
in Shakespeare's *Henry VI, Part II* (act 1, sc. 4): "The Duke yet
lives that Henry shall depose," she says, which leaves it unclear
whether "Henry shall depose the Duke" or "The Duke shall
depose Henry." (She could have made her meaning clear by sub-
stituting *who* or *whom* for *that.*) A much more interesting Shake-
spearian example, however, is the Witch's prophecy in *Macbeth:*

k)   Be bloody, bold and resolute; laugh to scorn
      The power of man, for none of woman born
      Shall harm Macbeth. (act 4, sc. 1)

"None of woman born" turns out to be a bitter deception when
Macbeth discovers that his enemy, Macduff, had had an unusual
birth, having been "ripped untimely from his mother's womb."
Macduff had apparently been born by Caesarian operation and so
was not "of woman born" in the usual sense. The Witch was therefore not lying to Macbeth, but she was not exactly telling him the
whole truth either.

No doubt the possibility of such ambiguity is why, in taking an
oath, we do not swear simply to "tell the truth" (for we could then
tell just part of it), nor even to "tell the truth, the whole truth" (for
we could then throw in also a few lies, not having sworn not to do
so), but rather "to tell the truth, the whole truth, and nothing but

the truth." Had the Witch sworn that oath, she could not have used the words she did to Macbeth.

Amphiboly has been used to deceive not only in fiction but in fact as well. The classic example concerns Croesus and the oracle at Delphi. Contemplating war with Persia, Croesus consulted the oracle regarding the outcome. He received the oracular pronouncement that "if Croesus went to war with Cyrus, he would destroy a mighty kingdom." Delighted with this prediction, Croesus went to war and was swiftly defeated. Upon complaining to the oracle, he received the reply that the pronouncement was correct: In going to war he *had* destroyed a mighty kingdom—his own! The account is given by Herodotus, the first Greek historian, who castigates Croesus for being so dumb:

> But as to the oracle that was given him, Croesus doth not right to complain concerning it. . . . It behooved him, if he would take right counsel, to send and ask whether the god spoke of Croesus' or Cyrus' kingdom. But he understood not that which was spoken, nor made further inquiry: wherefore now let him blame himself. (*The Histories,* bk. 1, ch. 91)

Besides these ancient or purely literary examples, cases have been recorded of lives and fortunes saved or lost as a result of ambiguities. An interesting and well-known example is the case of the Russian prisoner who sought release from a Siberian prison by appealing to the czar for a pardon. The czar returned the unpunctuated reply, "Pardon Impossible To Be Executed." He meant for the prisoner to be executed (the period intended after the word "Impossible"), but the jailor in charge read the message to mean "Pardon. Impossible To Be Executed" and released the prisoner. How the jailer fared subsequently is not related; no doubt when the error was discovered, he was substituted for the prisoner—judging by typical czarist justice.

Amphiboly is also capable of being exploited for purposes of gain. A typical example is the record entitled "Best of the Beatles," which misled many people into buying it, believing they were purchasing a record featuring the best songs of the Beatles. When they played it later at home, they discovered they had purchased a record featuring Mr. Peter Best, who had been a member of the Beatles (their drummer) early in the group's career.

2. THE FALLACY OF ACCENT

Unintended meanings can arise not only from faulty sentence structure, as in the case of amphiboly, but also from confusion as to emphasis. The *fallacy of accent** results when (1) a statement is spoken in a tone of voice not intended for it; (2) certain words in it are wrongly accented or stressed; or (3) certain words (or even whole sentences and paragraphs) are taken out of context and thus given an emphasis (and therefore a meaning) they were not meant to have.

> **Accent* was the term applied by Aristotle to misinterpretations resulting from words that differ in syllabic accent. An example in English would be the confusion of *in*valid (meaning "someone ill") and in*val*id (meaning "a faulty argument"). By extension the term came to be applied to whole words and sentences that when similarly misaccented convey a meaning they were not intended to convey.

The importance of this fallacy can be gathered from an incident surrounding the Watergate tapes. In a transcript of one of the tapes, John Dean, the president's counsel, warns Richard Nixon against getting involved in a cover-up and the president replies, "No—it is wrong, that's for sure." The question is: What inflection did Nixon give the words when he made this remark? Was it said in a serious and straightforward tone of voice, or was it said ironically? If it was uttered ironically, this remark would represent additional evidence of his involvement.

The fallacy of accent occurs less often in oral speech, where tone is easily conveyed by voice and gestural cues, than in written language. It is for this reason that clerks of the court deliberately read testimony aloud in a monotone, trying in this way to keep out any indications of their own feelings about the matter they are reading. To say in writing that "it is impossible to praise this book too highly" may mean either that we cannot say enough good things about it or that it is not worthy of praise at all. Similarly, to say that we hope someone "gets everything he (or she) deserves" could imply two very different meanings. Such confusions arise from ambiguities of tone, where it is necessary to know the attitude behind the language.

Questions of stress occur where the placement of an accent makes a difference in meaning. It is possible to ask the question, "Did you go to the store today?" with several meanings, depending on which word is stressed.

An old sea story illustrates this fallacy well. A sea captain and his first mate did not get along well, since the mate loved to drink and the captain did not. One day the captain thought he would fix the mate and entered in the ship's log the note, "The mate was

"I think you're over-reacting, Senator! ... Just because a voter writes and hopes you get what you deserve is not a threat on your life!"

Reprinted with special permission of NAS, Inc.

Accent is a simple fallacy which can be put to subtle use in more complex arguments.

drunk today." When the mate's turn came to keep the log he noticed the entry and was furious. The owner, he thought, would certainly fire him on his return to shore. To get revenge on the captain, the mate entered the note, "The captain was not drunk today," thereby implying that this state of affairs was so unusual that it merited mention in the log.

A command given at the Eucharist has given rise to questions of accent. Does "Drink ye all of it" mean that all of you should drink it, or that only some should drink it, but drink it all up? Dispute over this point has been widespread but has now been resolved in the more modern translation which reads, "Drink it, all of you."

The fallacy of accent also arises when whole sentences (and not merely one word or a few) are wrongly accented, as when we take such sentences out of their original context. This is a favorite device not only of propagandists but of some blurb writers and newspaper writers. Since no one has the time to check everything quoted in testimonials, newspapers, and other reports, the amount of misinformation conveyed by misleading captions, headlines, and quotations can be sizable. A drama critic might write, caustically, that she "liked all of the play except the acting," only to find herself quoted the next morning by an irresponsible advertising writer as having said she "liked all of the play. . . ." Persons reading this quotation out of context would be led to believe that what has been omitted from the quotation is an expansion of how favorably the play was received by this critic, when in fact the opposite is the case.

3. THE FALLACY OF HYPOSTATIZATION

Everyone probably remembers that delightful encounter between Alice and the cat in Lewis Carroll's *Alice in Wonderland.* On taking her leave, Alice sees the cat start to vanish slowly, "beginning," as she tells us, "with the end of the tail, and ending with the grin, which remained some time after the rest of it had gone." Alice is led to remark in astonishment: "Well! I've often seen a cat without a grin; but a grin without a cat! It's the most curious thing I ever saw in all my life!" (ch. 6).

If we were properly trained in logic, we would feel the same sense of astonishment that Alice felt whenever we heard anyone speak of *redness* or *roundness,* or *truth, beauty,* and *virtue*—as if they could exist by themselves and in their own right and were not merely abstractions, which (like the cat's grin) depend on some concrete entity for their existence. Imitating Alice, we would express our astonishment by remarking that we had certainly seen red *apples* or round *balls,* and truthful, beautiful, and virtuous *people,* but never roundness or redness, or truth, beauty, and virtue *as such.*

The *fallacy of hypostatization** consists in regarding an abstract word as if it were a concrete one. Whereas concrete words designate particular objects or attributes of objects, such as *red*

---

*Hypostatization comes from the Greek *hypostatos,* meaning "having an existence in substance." The Greek prefix *hypo* means "down" or "under" (as in *hypodermic,* "under the skin") and *statis* comes from the Greek root meaning "standing," which is related to Greek words for substance and sediment. To *hypostatize* in logic has thus come to signify the act of treating essence words as if they were substance words.

---

and *ball,* abstract words designate general qualities, such as *redness, roundness, virtue.* It is a peculiarity of abstract terms that they can be used without reference to subjects that possess the attributes they designate.

Although abstractions are a useful feature of language and thought, enabling us to discuss ideas like beauty or goodness, they carry potential dangers. We may make the mistake of assuming that, like concrete words, they name specific individual entities— that, for example, in addition to there being in the world such things as red balls and virtuous people, there are also separate entities such as redness, roundness, and virtue.

The process at work in hypostatization is similar to *personification.* To personify is to ascribe to things or animals properties that only human beings possess. It is to speak of things or creatures that are not persons as if they were persons. For example, we personify if we complain of the "cruelty of weasels" because weasels, being innocent creatures, cannot be considered either kind or cruel. To be cruel is to intend and plan some harm, knowing that it will cause pain, whereas weasels are not capable, as far as we know, of entertaining such designs. They simply are as they are and do as they do. The same applies to expressions such as "the cruel sea." Understood literally, this personification is simply false. To hypostatize is to speak of abstract entities in terms that are similarly appropriate only for persons.

It is thus that we may say, "The state can do no wrong," or "Science makes progress," or "Nature decrees what is right." Since no one of the three—the state, science, or nature—is capable of thought or intention, it is absurd to suppose that such abstractions are capable of the activities attributed to them in the statements above. Only persons, not the state, can be said to do right or wrong, only scientists can make progress, and nature has no voice with which to utter decrees.

To be sure, we do not usually lose sight of reality in most instances when we resort to hypostatization. When we say that "our budget dictated what we were to do," we are perfectly aware that we mean that our budgetary considerations—not some thing called *Budget*—compelled us to do what we did. We also realize that we are speaking metaphorically when we use such similarly innocuous examples of hypostatization as, "Love is blind," or "Facts call us now to bethink ourselves," or "Actions speak louder than words."

It is less certain, however, that we know what we are about when we say, "The city is aroused," or "The state is the march of God through history," or "It is the spirit of the nation which produces its art and literature." In such cases we may have been misled by our capacity for abstraction into thinking that, for example, there actually exists a grand artist at work whom we call the spirit of the nation.

As an indication of the dangers inherent in hypostatization, consider this argument:

a) Nature produces improvements in a race by eliminating the unfit and preventing them from polluting the gene pool of the fit. Therefore it is only right for us to eliminate these unfit people.

Nature is especially favored as a subject for hypostatization, perhaps because it is such a complicated abstraction that we have difficulty speaking of it at all without concretizing it. In argument *a* above, nature is endowed with an ability to know what is an

"For God's sake, can't you just take down the data it feeds us without exclaiming 'You're *so* right!'?"

The programmer employs the fallacy of personification when he speaks to the computer as if it were a person.

"improvement" and what is not, what is "fit" and what is "unfit," although it is unrealistic to impute to nature any humanlike intelligence or intention.

The tendency to hypostatize nature is doubtless related to many people's belief that nature was created by God and is therefore part of the divinely ordained order of things. Although ancient, this idea is the source of relatively more recent doctrines such as natural law. People differ as to whether the word *God* is abstract or concrete, but it is always helpful to know in which of these two senses it is being used on any given occasion. Argument *a* above would be significantly altered if the word *God* were substituted for *Nature*.

If talk of this sort is so harmful, why do we resort to it so frequently? The answer is simply that we get a lot of mileage from such talk. We use it, whether we realize it or not, as a form of verbal magic. Thus the term *science* used in an ad will sell almost anything, as advertisers have come to learn. We exploit the fallacy, too, of course. For example, we find it much easier to condemn "the establishment" or "the system" rather than to point out the particular laws or practices that we feel need changing.

But we come to pay a stiff price for this abuse. Take the term *freedom* used so often as a rallying call. We would all gain more were we to pause at such moments and ask ourselves: freedom for whom? freedom from what? freedom to do what? Such questions would save many lives.

We need to remember that terms such as *nature, truth,* and *freedom* are abstractions that do not exist in the real world. We can learn to depopulate that abstract world by a process of substitution. Whenever a term appears suspect, it should be traced back to the thing it seems to signify and a new term should be substituted for the suspect one. Sometimes the abstraction simply has to be changed from a noun to an adjective. Thus, we can replace "The truth shall make you free" with "Truthful statements shall make you free" and "The state can do no wrong" with "The president (*or* the members of Congress) can do no wrong." With this kind of analysis, many statements that seem deeply profound often prove not to be profound at all.

Despite what has been said, we ought to remember that not all uses of hypostatization are harmful or dangerous. In the context of poetry or literature this linguistic device is merely a kind of conceit that gives pleasure and harms no one. Here, too, the word *nature* seems to be especially favored. Shakespeare, for example, makes frequent use of it, as in *Julius Caesar,* where he has Antony say of Brutus:

> b) Nature might stand up
>   And say to all the world. "This was a man!"
>   (act 5, sc. 5)

And so does Alexander Pope in this couplet, written (c. 1730) and intended as an epitaph for Isaac Newton in Westminster Abbey:

> c) Nature, and Nature's laws lay hid in night
>   God said, "Let Newton be!" and all was light.

And Robert Burns in his poem "Let Not Women E'er Complain" (1794):

> d) Let not woman e'er complain of
>   Inconstancy in love!
>   Let not women e'er complain
>   Fickle man is apt to rove!
>   Look abroad thro' Nature's range,
>   Nature's mighty law is change!
>   Ladies, would it not be strange
>   Man should then a monster prove?

And Eddie Cantor, who had a running feud throughout his career with Georgie Jessel, requested that the following inscription be placed on his gravestone when he died:

> e) Here in Nature's arms I nestle
>   Free at last from Georgie Jessel!

One more, from a tombstone with a rather different sort of inscription, reads:

> f) Death is a debt
>   To Nature due
>   I have paid mine
>   And so must you.

Nor should we object to the use of this linguistic device in news reports and headlines where it is used for purposes of brevity, as in "Germany invaded Poland on September 1, 1939"; "Ottawa Rejects Treaty"; "Parliament in Session"; and so on. In such cases we understand clearly that it was the German army that invaded Poland on September 1, 1939; that it was the officials concerned, and not Ottawa as such, who rejected the treaty in question; that it was the members of Parliament, and not the abstraction *Parliament,* that met.

We must also recognize those cases in which we are not taken in by this form of speech but rather consciously use it with full awareness of what we are about. Christopher Stone's environmentalist work *Should Trees Have Standing? Toward Legal Rights for Natural Objects* is an excellent example of this.

For a very long time, as we all know, those who have tried to preserve the environment from abuse have not had an easy time of it—especially not with the arguments with which they have gone to court. Owners of forests and lakes have been able to make short shrift of the standard appeals: We are not to cut down our own forests because once a year you come driving through them? We are not to dispose of our industrial waste in our lakes because you like to fish in them? We need the pulp from those forests to produce the newspapers, magazines, and books which sustain our way of life, and we need our factories and their production to ensure jobs and goods for our people. And besides, the forest is our *property.*

What Stone offers in his little book is a radically different argument—and a far more powerful one. Put very plainly, what he says is this: Do not cut down that tree or pollute that lake, not because your action will spoil the pleasure others take in these beautiful, natural objects, but rather because the tree does not like being cut down and the lake does not like being polluted. In short, trees and lakes too have rights because they are beings in their own right and not simply the possessions or possible sources of pleasure for other beings. How egocentric it was of us, Stone argues, to assign to things moral significance purely on the basis of their use to us. We must try to reorient our attitude to nature, cease to view it as our toy or possession to do with whatever we like, and come to regard it as an entity having a value in itself worth protecting.

In his preface to this widely read and admired book, Stone tells the reader that if the view he is advancing (that natural objects have inalienable rights) appears absurd, it will be useful to remember that not very long ago such groups as women, African-Americans, and native Americans were also denied basic human rights on the ground that they were not quite human. Perhaps in time the argument that trees have rights will not seem as strange as it may at the moment.

What Stone is doing here is inventing a new fiction, pretending that trees are like persons (or, at least, like corporations) and have rights. He is, in short, hypostatizing or personifying natural objects. But, as Garret Hardin, in his foreword to Stone's book, points out, so do the owners of forests and lakes who have claimed those things as their "property":

> The most rigid defenders of the momentary legal definition
> of "property" apparently think "property" refers to some-

thing as substantive as atom and mass. But every good law-yer and every good economist knows that "property" is not a thing but merely a verbal announcement that certain tradi-tional powers and privileges of some members of society will be vigorously defended against attack by others. Operation-ally, the word "property" symbolizes a threat of action; it is a verb-like entity, but (being a noun) the word biases our thought toward the substantive we call *things.* But the per-manence enjoyed by property is not the permanence of an atom, but that of a promise (a most unsubstantial thing). Even after we become aware of the misdirection of attention enforced by the noun "property," we may still passively ac-quiesce to the inaccuracy of its continued use because a de-gree of social stability is needed to get the day-to-day work accomplished. But when it becomes painfully clear that the continued unthinking use of the word "property" is lead-ing to consequences that are obviously unjust and socially counterproductive, then we must stop short and ask our-selves how we want to re-define the rights of property. (*Should Trees Have Standing? Toward Legal Rights for Nat-ural Objects.* Los Altos, CA: William Kaufmann, Inc., 1974, pp. 6–7)

Hardin is not necessarily condemning the practice or process of hypostatizing he describes. He is merely urging us to recognize it for what it is. "Property" is a fiction, a notion we have devised or invented. It doesn't stand for some concrete thing we discover that has an existence independent of us or of our wishes. As long as we find the concept useful and helpful we can continue to embrace it; if the notion starts to prove otherwise, let us remember that we ourselves brought it into being (defined it), and nothing therefore should prevent us from banishing it (redefining it) should we de-cide to do so.

Of course, we are not likely ordinarily to meet the fallacy in this unusually tricky and elegant form. More likely we will find our-selves manipulated by a statement such as, "The world will no longer laugh," never asking ourselves how many of the 4 billion people in the world indeed know us, and of those few who do, how many care one way or another what we do. British Lord Nelson knew the force of such a slogan: Before the Battle of Trafalgar (October 1805) he gave the naval order, "England expects every man to do his duty!" Notice, he did not say *he* expects it (which would have been intimidating enough, for he was an imposing figure) but that *England* does: He meant every English person, all English history and tradition, its honor, and so on. Now that should be sufficient to make a man lay down his life!

Even if we are not psychologically subdued or seduced by slogans beginning with "The world" or "People," we may still find ourselves deceived in the manner of the following letter writer:

> g) What Americans need is patience. A water well goes dry. Nature replenishes it by sending a good rain. Nature shall replenish the great oil reservoirs. Scientists say it takes a few million years. All we need is patience.

"Nature" is not a person; therefore, it is doubtful that it is aware that our wells are going dry and need "replenishing," to say nothing of the absurdity of believing we can wait "millions of years" for it to happen.

We must beware of such grand concepts, or we may find ourselves haunted by hypostatized specters, as is the following writer:

> h) Because the dumping ground was so far out to sea, the city believed it would never hear from its sludge again, but the city was wrong. The mass of goo slowly grew and sometime around 1970 it began to move, oozing back to haunt New York and the beaches of Long Island.

4. THE FALLACY OF EQUIVOCATION

Thus far we have seen how ambiguity of sentence structure can give rise to confusion in the form of the fallacy of amphiboly; how ambiguity of stress or tone can bring about the fallacy of accent; and, finally, how ambiguity in the reference made can result in the fallacy of hypostatization. We shall now consider how confusion can arise from the ambiguity of words or phrases themselves.

To commit the *fallacy of equivocation** is to allow a key word in an argument to shift its meaning in the course of the argument.

Equivocation is from the Latin for, literally, "equal" *(equi)* "voice" *(vox)*. A word is used *univocally* if it has the same meaning throughout a given context, *equivocally* if one or more other meanings are equally possible.

Consider this example:

a) Only man is rational.
    No woman is a man.
    Therefore no woman is rational.

This argument would be valid if the word *man* had the same meaning each time it occurred. But for the first premise to be true, *man* must mean "human," whereas it must mean "male" for the second premise to be true.

When the change in meaning of a key word during an argument is especially subtle, the conclusion will seem to follow clearly from the premises and the argument will appear considerably more sound than it is.

b) The financial page of the London *Times* says that money is more plentiful in London today than it was yesterday. This must be a mistake, for there is no more money in London today than there was yesterday.

In the context of this argument, the terms *plentiful* and *more* at first seem equivalent. On closer examination, however, the first term is seen to refer to the distribution of money, the second to amount.

The fallacy of equivocation is especially easy to commit when a key term in an argument happens to be a figure of speech or a metaphor. By interpreting the metaphor literally we sometimes persuade ourselves that an argument is sounder than it is. Few are likely to be seriously misled by a figure of speech such as "he has a lean and hungry look," and our language would be poorer without such expressions. But many figurative expressions need to be used with caution.

c)   It is the clear duty of the press to publish such news as it shall
be in the public interest to have published. There can be no
doubt about the public interest taken in the brutal murder of
Countess Clamavori and concerning the details of her private life that led up to the murder. The press would have
failed in its duty if it had refrained from publishing these
matters.

Here the expression *the public interest* means "the public welfare" in the first premise, but it means "what the public is interested in" in the second premise. Thus the argument is fallacious
because what the public is interested in is not the same as what
is in its best interest.

The following argument is similar to argument *c* above:

d)   No one who has the slightest acquaintance with science can
reasonably doubt that the miracles in the Bible actually took
place. Every year we witness new miracles of modern science such as television, jet planes, antibiotics, heart trans-
plant operations, heat-resistant plastics.

Which is the metaphorical use of *miracle* and which the literal?
What does the expression *miracles of science* mean? A similar
shift in meaning characterizes the next argument.

e)   As far as I'm concerned, we need pay no attention to the president of the college when it comes to educational matters
because he has no authority in education. He doesn't even
have enough authority to prevent students from staging protest rallies.

The inference is fallacious because the arguer equivocates.

Equivocation is not confined to figurative expressions, for the
vast majority of our words have more than one meaning, any of
which can occasion the fallacy. Many delightful examples are
found in that rich source of verbal nonsense, Lewis Carroll's
Through the Looking Glass.

f)   "Who did you pass on the road," the King went on, holding out
his hand to the Messenger for some hay.
"Nobody," said the Messenger.
"Quite right," said the King, "this lady saw him too. So of course
Nobody walks slower than you."

"Here, you wanted to be plant manager. Take care of this!"

Exploiting equivocation.

> "I do my best," the Messenger said in a sullen tone. "I'm sure nobody walks much faster than I do!"
>
> "He can't do that," said the King, "or else he'd have been here first!" (ch. 7)

A mathematician in Victorian England and the author of a book on logic, Carroll wove into his stories many paradoxes of language and of logical thought.

Among the words with multiple meanings which are particularly susceptible to equivocation are those whose meanings are vague and indeterminate. The word *practice* in the following example is of this type.

> g)   Practice makes perfect. Physicians have practiced the art of healing for thousands of years. My physician, therefore, who

studied at one of our greatest medical schools, should be perfect in his field.

On an intuitive level, this argument is easily recognizable as unsound because my physician cannot be equated with all the physicians who ever lived; he himself has certainly not studied medicine for thousands of years so as to become perfect. Viewing the argument as an example of equivocation, on the other hand, enables us to see its flaws more clearly. On this level we can recognize a failure to distinguish between what can be called the practice of medicine, which as a profession has existed for thousands of years, and one physician's practicing medicine. Furthermore, we can recognize that *practice* is used in yet another sense in the premise, "Practice makes perfect."

Such analysis also enables us to see why arguments containing fallacies of equivocation often appear plausible. Because the premises out of which they are composed are all unobjectionable when considered individually, any shift in meaning from one statement to another may escape our notice. Both of the premises in argument *e* above are plausible. When they are put together to form an argument, the argument will seem plausible unless the reader recognizes that a key term has shifted in meaning as it proceeds from the one premise to the other. This is in fact the best method for testing the soundness of such arguments. If you suspect that a key term has shifted its meaning, simply reread the argument, keeping the meaning of the suspect term uniform. In many cases, this technique will reveal either an absurd premise or an absurd conclusion.

It is useful to remember that at the root of the vast majority of cases of equivocation lies the appeal that we not contradict ourselves, that we be consistent. The drive to be consistent can be a trap, however. Consider this example going back several centuries:

h) There are laws of nature.
    Law implies a lawgiver.
    Therefore there must be a cosmic lawgiver.

Our reply in this case should be: since *laws* as used in the context of "laws of nature" simply means "set of observations" and *law* as used in the context of "lawgiver" means "set of commands," we are not contradicting ourselves in believing in the one and not in the other.

Similarly instructive is this somewhat longer example:

i) In our democracy all people are equal. The Declaration of Independence states this clearly and unequivocally. But we tend to forget this great truth. Our society accepts the principle of competition. And competition implies that some people are better than others. But this implication is false. The private is just as good as the general; the file clerk is just as good as the corporation executive; the scholar is no better than the dunce; the philosopher is no better than the fool. We are all born equal.

Our reply: The fact that we believe that all of us have the same *rights* does not mean we must also believe what is obviously false, that all of us have the same *abilities* (and therefore should be treated "equally").

Here is a still trickier example:

j) I do not believe in the possibility of eliminating the desire to fight from humankind because an organism without fight is dead or moribund. Life consists of tensions. There must be a balance of opposite polarities to make a personality, a nation, a world, or a cosmic system.

Our reply: Since the phrase *desire to fight* means essentially "violence," and the phrase *without fight* means "will," "drive," or "spirit," it may very well be possible to eliminate the one (violence) without necessarily destroying the other (our spirit).

Finally, consider this example from an editorial:

k) I am puzzled by the protest groups that gather in front of prisons when an execution is scheduled. The murderer, who has committed a heinous crime, has been granted all due process of law and is given every opportunity to defend himself— usually with the best available legal minds and often at taxpayers' expense. Yet these same protestors generally favor the execution of millions of innocent babies by abortion.

Our reply: We are not being inconsistent in protesting the *execution* of people found guilty of murder and not the *execution* "of millions of innocent babies by abortion." In the one case it is indeed an "execution" (for the life of a person is definitely involved) and not in the other (for a fetus is arguably not yet a person and perhaps not yet even a "life").

We see from the examples just noted how very intimately the words we use to express ourselves enter into the thinking represented in the arguments. It is as though the words themselves actively lead us astray.

A particularly striking example of the direct influence of language on thinking is the famous fallacy committed by the English philosopher John Stuart Mill in one of his works on ethics. Mill is here dealing with the question of what is the most *desirable* end or aim of human conduct and argues that it is happiness, as utilitarianism teaches. But how can we prove that happiness is indeed the one true ideal that we should desire? To this Mill replies:

> The only proof capable of being given that an object is visible, is that people actually see it. The only proof that a sound is audible, is that people hear it: and so of the other sources of our experience. In like manner, I apprehend, the sole evidence it is possible to produce that anything is desirable, is that people actually desire it. (*Utilitarianism,* ch. 4)

But critics of Mill have pointed out that he had been deceived by his manner of expressing himself. Though the words *desirable, visible,* and *audible* are structurally similar, they are not semantically the same. *Desirable* is not related to *desired* in the same way that *visible* and *audible* are related to *seen* and *heard,* for the first involves a moral distinction that the other two do not. *Visible* means simply that something is "capable of being seen" (and *audible* means something "capable of being heard"), but *desirable* implies that something is "worthy of being desired," that it "ought" to be desired. This being so, it may be quite true that a thing's being seen or heard proves that it is visible or audible, but it does not follow that, because a thing is desired, it is for that reason desirable. Many people may desire drugs, but that does not prove that drugs are therefore desirable.

Mill had unfortunately been taken in by what we might call a systematically misleading expression. He made the same error involving *desirable* that we would if we were to note that since, say, *immaterial* means "nonmaterial" and *insoluble* means "nonsoluble," therefore *inflammable* must mean "nonflammable" (when it means just the opposite of that). Another example of this type of error is that, since *decompress* means to lower or remove compression and *decentralize* means to remove centralization, therefore *delimit* must mean "remove limits" (when, again, it means the very opposite, namely, to "impose limits"). Obviously, words similar in structure can be very different in meaning.

5. THE FALLACIES OF DIVISION AND COMPOSITION

If a person cannot break a bundle of sticks, does that mean he or she cannot break any one of them individually? Of course not.

What is true of the whole is not necessarily true of its parts. To think so is to commit the *fallacy of division*. It is to try to divide what is true of the whole among its parts.

We can reverse the order of the argument and arrive at the opposite fallacy: If a person can break one stick, then another one, then still another one, does that mean that individual can break the bundle of sticks as a whole? Probably not.

What is true of the part is not necessarily true of the whole. To think so is to commit the *fallacy of composition*. It is to compose the whole out of its parts. The whole, as the old saying has it, is more than the sum of its parts.

What is true here of wholes and parts is also true of groups and their members. Thus the Chicago Symphony Orchestra may be the best orchestra in the country, but that does not necessarily mean that the first violinist in the orchestra is the best violinist in the country. Or to use another example, Jones may be the best quarterback in the country and Smith the best halfback and Davis the best receiver—but putting them and other outstanding players together on one team will not (as we have seen only too often) necessarily give us the best team in the country.

Why, one might ask, is it not the case that what is true of the whole is not necessarily true of the parts? Or why is it not the case that what is true of the parts is not necessarily true of the whole? Or, similarly, why is it not the case that what is true of some particular group or team is not necessarily true of the members of the group or team, and vice versa? The reason is that a whole or group is something functional and organic and therefore has properties just in virtue of being such a whole or group.

Another example will make this clearer: A woman looks at a flower and says to herself, "Oh, what a very pretty flower!" She does the same with several others. Does this mean that if she gathers them together, she will therefore have a pretty bouquet of flowers? Perhaps not, and the reason is that bringing them to-

"I judge a man by the shoes he wears, Jerry."

"The Far Side" cartoon by Gary Larson is reprinted courtesy of Chronicle Features, San Francisco.

In this case, the woman commits the fallacy of composition. What is true about the man's shoes might not be true about the man as a whole person.

gether gives rise to something new. Will all the different flowers blend properly? That question was not pertinent when each flower was considered by itself. The same would apply to a group of players, or a group of singers, or any such collection of individuals. Although each person, when considered by himself or herself, might be outstanding, whether the group will be outstanding depends on a new factor that arises only with the formation of the group. How well will they work together or how well will their different voices blend together? These are questions that had no meaning when each member of the group was considered individually.

Although they are obviously not very difficult or subtle kinds of fallacies, we tend to commit them rather frequently. Often the reason lies in a confusion of the *collective* and the *distributive* sense of certain key terms in arguments. By a collective term we mean a term that refers to a collection or a whole; by a distributive term we mean a term that applies only to individuals or parts. The word *all* is probably the best example of such a potentially ambiguous term. When we say, for example, "All donors have contributed $1,000" do we mean that each and every one of them has contributed this amount (using the word *all* distributively), or do we mean that all, taken together, have done so (understanding the word *all* now collectively)?

A final word of caution. Logicians remind us that it is a fallacy to *presume* without further evidence that a whole will have the properties possessed by each of its parts, or that whatever is true of any member of a group is true of every member. This does not mean that we may not occasionally run across cases where what is true of the whole is indeed true of its parts, or vice versa. The rule states that we may not presume that it will always be so. Thus if the truck you bought recently is brand new, it is likely the parts are so too. (But even here manufacturers have been known who have used old parts when they ran out of new ones.)

In addition we must remember that these remarks about the relations between wholes and parts and groups and members concern physical wholes and their parts. Where the wholes and parts or the groups and their members are other than physical, no fallacy may be involved.

Roger Rosenblatt's essay (*Time,* December 17, 1984) in response to the disastrous Union Carbide poison-gas accident in Bhopal,

India, which took the lives of more than 2,500 people is a striking case in point.

Rosenblatt begins his essay by quoting John Donne's famous line that "Any man's death diminishes me" (which, on the surface, may appear to commit the fallacy of division). He says of it:

> It has always sounded excessive. John Donne expressed that thought more than 350 years ago in a world without mass communications, where a person's death was signaled by a church bell. "It tolls for thee," he said. Does it really? Logic would suggest that an individual's death would not diminish but rather enhance everybody's life, since the more who die off, the more space and materials there will be for those who remain. Before his conversion, Uncle Scrooge preferred to let the poor die "and decrease the surplus population." Scrooge may not have had God on his side, but his arithmetic was impeccable.

Was John Donne guilty then of this fallacy of division?

> Are Donne's words merely a "right" thing to say, then, a slice of holy claptrap dished out at the Christmas season? What does it mean to believe that any man's death diminishes me? In what sense, diminishes? And even if one wholeheartedly accepted Donne's idea, what then? What use could one possibly make of so complete an act of sympathy, particularly when apprised of the deaths of total strangers?

Donne would have been guilty of the fallacy had he in mind only the physical results of death.

> But Donne seemed to be advocating a response that is deeper and more consistent. Any man's death makes me smaller, less than I was before I learned of that death, because the world is a map of interconnections. As the world decreases in size, so must each of its parts. Donne's math works too. Since the entire world suffers a numerical loss at an individual's death, then one must feel connected to the entire world to feel the subtraction equally.

Rosenblatt's last point is that one comes to feel so connected with the rest of the world and humankind as a whole by realizing our shared vulnerability in the face of the arbitrariness of fate:

The world can feel as small as Donne's. If nothing else, we have vulnerability to share. A reporter walking about Bhopal last week remarked how on some streets people were living normally, while adjacent streets were strewn with bodies. Everything depended on where the wind was blowing.

SUMMARY

This chapter has presented six fallacies of ambiguity. Such fallacies were shown to be linguistic fallacies, in that they stem from use of language having more than one meaning. We saw that the best way to unravel such fallacies is to clarify the language in question.

Amphiboly was shown to result from ambiguity in sentence structure, as when Croesus draws the wrong conclusion from the oracle who prophesies, "If Croesus went to war with Cyrus, he would destroy a mighty kingdom."

Accent was seen to arise where there is ambiguity of stress or tone, as in the case of the senator who makes a fallacious inference from the letter from a voter saying he "hopes the senator gets what he deserves."

Hypostatization was shown to result when an abstract word or phrase is used as if it referred to something concrete. Such an abstraction was exemplified in the argument claiming that, because nature improves a race by eliminating the unfit, it is right for one group of people to eliminate another group.

Equivocation was the name given to fallacies stemming from a shift in meaning of a key term during an argument. When we argue that "only man is rational" and that, because women are not men, it follows that "no woman is rational," we change the meaning of the word *man* during the course of the argument.

Division was shown to result from trying to apply what is true of the whole or the group to each part or member. But, as we saw, the Chicago Symphony Orchestra may be the best orchestra in the country, but that does not necessarily mean that the first violinist in the orchestra is the best violinist in the country.

Composition was shown to result from trying to apply what is true of the part or the individual to the whole or group. But, as we saw, a bundle of sticks or football team is more than merely the sum of its parts or members.

EXERCISES

Identify the fallacy of ambiguity—amphiboly, accent, hypostatization, equivocation, division, or composition—which is committed in or which could result from each of the following. Explain the error committed in each case.

* 1. I shall lose no time in reading your paper.

2. The verdict rendered by the jury was reasonable and just. Mr. Black, who was a member of the jury, is therefore a reasonable and just man.

* 3. Tourists Taken In (Road sign)

* 4. Our X-ray unit will give you an examination for tuberculosis and other diseases which you will receive free of charge. (Public service announcement)

5. All I keep hearing from Mom and Dad is the importance of nutrition. Why do I have to eat carrots, spinach, and liver? Why don't they just give me some nutrition to eat?

* 6. We stand behind every bed we sell.

* 7. You say nothing eloquently.

8. No one on this committee is especially outstanding in ability. It is impossible for the committee, therefore, to bring in an able report.

* 9. The end of a thing is its perfection; death is the end of life; death is, therefore, the perfection of life.

10. The fund has a deficit of $57,000, which will be used to pay teachers' salaries.

*11. The world does not owe you a living.

12. We are told that prostitution is a growing national problem, but that isn't the half of it! At least half of the men and women in this country today are prostitutes. They sell their bodies or their minds in jobs that are personally meaningless and socially destructive.

13. If you can tie a knot, you can make a beautiful deep pile rug! (Advertisement)

14. Come to us for unwanted pregnancies. (Billboard)

15. Thou shalt not bear false witness against thy neighbor.

*16. We ought to be guided by the decision of our ancestors, for old age is wiser than youth.

17. Love your neighbor.

* 18.   It is impossible to be too enthusiastic about the new exhibition of sculpture.

* 19.   You have believed a lie that you cannot get well. The truth will make you free. Love Nature. She is gentle and holy. To obey her is to live. (Circular on healing)

* 20.   A different perspective was offered by Dewain Woljen, a former Corps of Engineers officer who has since become an ardent member of the committee against building the dam. "The corps is a bureaucratic entity," says Woljen. "As such, it has to constantly assert itself by doing what it was created to do—build. That's the way it retains its own sense of power. One doesn't have to accuse the corps of being in anybody's pocket to suggest that its interior needs and the needs of the developers often coincide." (News report)

21.   Easy new way to weld with instruction book. (Welding kit)

22.   Slow Children Crossing. (Street sign)

23.   "You are inconsistent," protested a member of the jury to the foreperson. "You told me yesterday that there was a presumption of this man's innocence, and now when I say that we may presume he is innocent, you contradict me!"

* 24.   You may think as you please.

* 25.   Mr. Raymond has accepted the post of organist. Extension of the graveyard has become necessary a year before expected. (Church newsletter)

26.   When a Kodak service representative visits a Kodak copier, the copier's powerful microprocessor does tell where it hurts. And where it's hurt in the past—a complete service history. And then it aids in diagnosis, correction and testing. All of which tends to keep service calls for Kodak copiers uniquely brief. Indeed many service calls don't even happen, because the computer's made an adjustment all by itself! Or told an operator how to do it. Now you know why Kodak copiers constantly outrank all others in surveys of copier service satisfaction. And why you should see a Kodak copier demonstration. (Advertisement)

* 27.   If Americans can be divorced for "incompatibility," I cannot conceive why they are not all divorced. I have known many happy marriages, but never a compatible one. For a man and a woman, as such, are incompatible. (G. K. Chesterton)

28.   Jones thinks McIntosh will succeed.

29. Why, if thou never wast in court, thou never sawest good manners; if thou never sawest good manners, then thy manners must be wicked; and wickedness is sin, and sin is damnation. Thou art in a parlous state, shepherd. (Touchstone, in Shakespeare's *As You Like It*)

* 30. You can skate more than one mile on one slice of bread. (Baker's advertisement)

* 31. It is enterprise which builds and improves the world's possessions. If enterprise is afoot, wealth accumulates whatever may be happening to Thrift; and if enterprise is asleep, wealth decays whatever Thrift may be doing. (John Maynard Keynes, *Treatise on Money*)

* 32. Abraham Lincoln was obviously not the man it is thought he was, for he said, "You can fool all the people some of the time."

33. The Marines are looking for a few good men. (Marine poster)

34. We do not tear your clothes by machinery; we do it carefully by hand. (Commercial)

* 35. It is not going to help ease the energy crisis to have people ride buses instead of cars. Buses use more gas than cars.

36. I hope you will stay for lunch.

37. Lost: Samsonite Briefcase with Eyeglasses

38. I've got nothing to say and I'll only say it once. (Floyd Smith, Toronto Maple Leafs coach, after a loss)

39. A National Theatre is worth having for the sake of the National Soul. (George Bernard Shaw)

40. Elderly Often Burn Victims (Headline)

41. The place is so crowded that nobody goes there anymore. (Overheard)

42. To enact reasonable drug legislation, we must find out about the effects of the various drugs—whether they are habit forming, have had bad effects on hormones, lead to psychoses, and so forth. For this reason we need the advice of an authority, and so I invited the chief of the police department to testify before the committee.

* 43. If you think our waitresses are rude, you should see our manager. (Sign in restaurant)

44. What is hell? Come to church next Sunday and listen to our new minister! (Church announcement)

*45. What can you say about a girl who wears Heaven Scent? (Advertisement)

46. We have the finest politicians money can buy. (Talk show)

47. I cannot get sick pay. I have six children. Can you tell me why? (Letter)

48. It is predicted that the consumer's price index will rise again next month. Consequently, you can expect to pay more for butter and eggs next month.

*49. A mob is no worse than the individuals who make it up.

50. Don't fear. Nature always looks after her offspring.

*51. Since the city revenues have fallen off, I propose a 20 percent across-the-board cut for all city departments. We'll just have to get along with four-fifths of the service we've been used to.

52. Each manufacturer is perfectly free to set his or her own price on the product he or she produces, so there can be nothing wrong with all manufacturers getting together to fix the prices of the articles made by all of them.

53. Flattering women amused him. (Student essay)

54. She didn't try to commit suicide today.

55. Accidents are frequent; getting struck by lightning is an accident; therefore, getting struck by lightning is frequent.

56. But what are these attackers of patriotism really against? They rise up in fury if anyone suggests they are opposed to patriotism itself; they simply say they dislike "too much" of it. They dislike "extremists" and prefer "moderates." These words and ideas have no sensible content. If patriotism is good, then a lot of it should be extremely good and who would not want something that is extremely good?

57. Life wants to preserve itself. Life wants to spread. Life wants to satisfy itself. Life wants to refine itself.

58. How can you doubt that the soul survives the body? Why, consider how each spring nature has a new birth in which creatures that were dormant during the winter come back to life again in renewed activity. Surely you cannot doubt that what is true of the rest of nature is true of people? You must agree, therefore, that they too are reborn—with life and soul entering into new life in a manner no less remarkable than the rebirth of flowers in spring.

59. If you are one of the growing number of Americans who realizes that American Telephone & Telegraph Company's slogan means "What's good for big business is good for America," then . . .

   Welcome to the *Progressive,* the monthly magazine that knows it's long past time to make fundamental changes. More and more of us see that:

   The System squanders our nation's wealth.
   The System rapes our natural and human environments.
   The System pours hundreds of billions of dollars down a rathole called "national security."
   The System puts profits ahead of people.
   The System works all right—it works for AT&T and Lockheed, for IBM and Exxon—but it doesn't work for us, the American people.

60. Volkswagen does it again. (Advertisement)

61. Every state is a community of some kind, and every community is established with a view to some good, the state or political community, which is the highest of them all, and which embraces all the rest, aims at a good in a greater degree than any other, and at the highest good. (Aristotle, *Politics*)

62. Can the universe think about itself? We know that at least one part of it can: we ourselves. Is it not reasonable to conclude the whole can?

63. If you own a young male collie or German shepherd trained for hunting and you are home on a summer afternoon, beware of being bitten by your dog. That is the statistical profile of the most likely conditions for dog bites, published in the November–December issue of *Public Health Reports.* Data collected from the study at Scott Air Force Base in Illinois and Whiteman Air Force Base in Missouri showed that collies and German shepherds were twice as likely to be biters as were mixed-breed dogs, and that working or sporting dogs are more likely to bite humans than are house pets. About half the bites occurred between noon and 6 P.M. and most took place during the late spring and summer months. (News report)

Find examples from your daily reading of the fallacies discussed in this chapter and explain why you think they are fallacious.

ANSWERS TO STARRED EXERCISES

1. *Accent:* Depending on the tone of voice in which this is spoken, it could mean the person will not waste any of his or her time in reading the paper (because it is such a bad one), or that he or she will not waste a moment in starting to read it (because it is such a fine one).

3. *Amphiboly:* Is the sign merely telling us that tourists are welcome there, or is it also letting us know that tourists are cheated there ("taken in")?

4. *Amphiboly:* What will be given free of charge—the examination alone, or the examination plus the diseases?

6. *Amphiboly:* Although this sounds as if they intend to place themselves physically behind each bed they sell, what they undoubtedly mean is that they guarantee the quality of each one.

7. *Accent:* Depending on which word is accented, this could mean: (a) *"you* say nothing eloquently" (though someone else does); (b) "you say *nothing* eloquently" (meaning there isn't anything you say well); (c) "you say nothing *eloquently*" (meaning even when you say nothing, you say it eloquently).

9. *Equivocation:* In the first premise *end* is used in the sense of "goal" or "purpose," meaning the reason why we do a thing; in the second premise *end* is used in the sense of "ending" or "finish." It does not follow, therefore, that since death ends life, it is life's goal.

11. *Hypostatization:* The world is not a human being aware of our needs or capable of taking on or fulfilling obligations. It cannot, therefore, owe (or not owe) us anything. What is intended is that each individual is responsible for his or her own life—whether it be earning a living or filling one's needs—and should not expect others to be responsible for those things.

16. *Equivocation:* Basing itself on the popular maxim that "old age is wiser than youth," the argument then goes on to conclude that we should therefore let ourselves be guided by "the decision of our ancestors" (that is, those living in former "ages"). To argue so, however, is merely to equivocate with the term *age*: it may be wise indeed to listen to the opinion and advice of older people in our own age, who having lived longer and having had more experi-

ence than us, may indeed be wiser than us, but that may not be true of those who lived in former ages whose lives and problems may have been too different from ours to be of similar value.

18. *Accent:* Depending on how this is spoken, the sentence could mean either that (a) it is impossible to praise this exhibit too much (it is such a fine one), or (b) it is impossible to praise it at all (because it is such a poor one). What, in addition, makes it easy to convey these two separate meanings here is that the word *too* can mean "excessively," and it can mean "very."

19. *Hypostatization:* Nature is a construct and not a person capable of gentleness and saintliness, or of giving orders. The probable intent of the remark, however, is that if one were to live a more simple life and were to reestablish a system of personal values, which gave more priority to spiritual health, one's body would become healthier.

20. *Hypostatization:* In this remark humanlike characteristics are being attributed to an abstract entity that cannot have them. A "corps" is not the sort of thing which "can assert itself," be "in [someone's] pocket," have a "sense of power," "interior needs," etc. Only persons can have such characteristics, and such descriptions can be attributed, therefore, only to them.

24. *Accent:* Depending on where the accent falls, this could mean: (a) *"you* may think as you please" (but not someone else); (b) "you may *think* (but had better not do!) as you please"; or (c) "you may think as *you* please" (but you're wrong anyway).

25. *Amphiboly:* Putting these two sentences together, as here, seems to imply that Mr. Raymond (or his playing?) has been responsible for the need to extend the graveyard.

27. *Equivocation: Incompatible* as used to refer to marriages means "being different temperamentally"; as used to refer to men and women ("as such"), it means "being biologically different, of different sexes." The argument is therefore unsound, for the fact that men and women are incompatible in being sexually different does not mean they are or will be similarly incompatible temperamentally, and thus never able to get along with one another.

30. *Amphiboly:* Because of the way this is worded, it seems to suggest that one can skate more than one mile *on the sur-*

face of one slice of bread; whereas what is probably intended is that one can skate more than one mile *on the energy* supplied by one slice of bread.

31. *Hypostatization:* Enterprise, being an abstract entity and not a human being, is not capable of knowing what would improve the world's possessions, nor does it have feet to walk on or a brain which might require sleep. The same applies to wealth and thrift. They too are only abstractions and not beings with awareness.

32. *Accent:* Lincoln is being quoted out of context. What he reputedly said was, "You can fool all the people some of the time, and some of the people all of the time, but you cannot fool all the people all of the time." In context, then, Lincoln's remark carries just the opposite implication here attributed to him.

35. *Composition:* Certainly each bus uses more gas than each car. Since, however, there are so many more cars than buses, it would indeed help ease the energy crisis if more people rode buses. The error lies in failing to see that what is true of *each* bus (as compared to each car) is not true of the *whole* group of buses (as compared to the whole group of cars); the amount of gas used by all cars put together is obviously vastly greater than that used by all buses put together.

43. *Accent:* Stress the word *manager* and it is a straightforward remark advising the person to see the manager about the complaint; on the other hand, stress *waitresses* and the remark is ironic, informing the person that in that restaurant when it comes to rudeness it is the manager who takes the prize.

45. *Accent:* Said enthusiastically, it compliments the person who wears that particular perfume; said sarcastically, it wonders what sort of weird person it is who would do so.

49. *Composition:* A group of intelligent people does not always act intelligently, nor does a group of civilized people always act in a civilized manner. A mob is not simply a large group of individuals. Something happens to people when they become parts of crowds.

51. *Division:* The argument is that since there is now 20 percent less money to distribute than formerly, what should be done is to give each department 20 percent less than it had been given previously, and the city, in turn, will sim-

ply have to get along with 20 percent less service that it will now get from each. The results, however, may be quite different: giving, say, the fire department 20 percent less money than formerly may result not merely in 20 percent less service from it but in no service at all, since that department may not be able to function with such a drastic cut. On the other hand, in other departments no reduction in service may result at all, and in some (perhaps from employee fear of job loss) the public may get better service than ever before.

"Wait a minute here, Mr. Crumbley . . . Maybe it isn't kidney stones after all."

4

Fallacies of Presumption

Fallacies of presumption are arguments that are unsound because of unfounded or unproven assumptions embedded in them. By smuggling such presumptions in under the guise of valid argument patterns, these fallacies give the impression of being like the valid arguments they resemble. However, since no conclusions can be more reliable than the assumptions on which they are based, the conclusions in such arguments cannot be trusted.

In fallacies of presumption, facts relevant to the argument have not been represented correctly in the premises. This inappropriate treatment of facts can take three forms. We may overlook significant facts entirely (as might have been done when the doctor in the cartoon at left made his original diagnosis), we may evade them, or we may distort them. This chapter will present the three types of presumptive errors and will examine the fallacies most common to each.

OVERLOOKING THE FACTS

In this first group of presumptive fallacies, the error committed is one of neglecting important facts relevant to the argument. In the

119

Table of Fallacies of Presumption

Fallacy	Definition/Hints	Example/Method
Sweeping Generalization	Applying a fair generalization, one usually true, to an exceptional case by ignoring the peculiarities of the case (Isolate clearly the generalization in the argument, and then see whether it was meant to be applied to the case in question.)	"Since horseback riding is healthful exercise, Harry Brown ought to do more of it because it will be good for his heart condition." (Because of his condition it may be dangerous for Harry to exercise, and he may have been instructed by his physician not to.)
Hasty Generalization	Using insufficient evidence or an isolated example as the basis for a widely general conclusion	"I had a bad time with my former husband. From that I've learned that all men are no good."
Bifurcation	Considering a distinction or classification exclusive and exhaustive when other alternatives exist (The other names for the fallacy are instructive: either/or fallacy; black-and-white fallacy; false dilemma.)	"You're either for me or against me!"

Table of Fallacies of Presumption *(Continued)*

Fallacy	Definition/Hints	Example/Method
Begging the Question (three forms)	(1) Offering, as a premise, a simple restatement of the desired conclusion (A because of B, where B is the *same* as A)	"Miracles are impossible (A) because they cannot happen (B)."
	(2) A circular argument, more complex than (1) but eventually justifying the conclusion with itself (A because of B, where B is *dependent* on A)	"GOD EXISTS (A)." "How do you know?" "THE BIBLE SAYS SO (B)." "How do you know what the Bible says is so?" "IT'S THE WORD OF GOD (A)."
	(3) Subsuming a suspect particular under a generalization that is even more problematic (A because of B, where B is even *more suspect* than A)	"Clearly he's an atheist (A); he's a philosopher isn't he (B)?" (Wider generalization: all philosophers are atheists)
Question-Begging Epithets	Using strongly emotional language to force home an otherwise unsupported conclusion	"The scheming, bigoted efforts of the Board of Education have finally come to fruition."
	(Often the epithet is critical—dyslogistic—but it can be laudatory—eulogistic.)	"Surely there can be no question Ed's our man. He's just a great guy."
Complex Question	The interrogative form of begging the question (the purported question bringing an assumption with it that needs to be questioned)	"Why is it that women are more interested in religion than men?"
	(Do not call an example a complex question unless it has a question mark in it.)	(Ask yourself: What does the question assume to be so? Is it really so?)
Special Pleading	Applying a double standard that is exemplified in the choice of words	"Horses sweat, men perspire, women glow."

Table of Fallacies of Presumption (*Continued*)

Fallacy	Definition/Hints	Example/Method
False Analogy	Reaching a conclusion by likening or comparing two significantly incomparable cases (Typically the two cases used will be similar but not in the respect that would warrant the conclusion in question.)	"How can you tell your children not to take money from others when the government they live under does it all the time?" (Point out how the two cases being compared resemble each other only in trifling ways and differ in profound ones.)
False Cause	Inferring a causal link between two events when no such causal connection has been established (Sequence alone is no proof of consequence.)	"Have you noticed how the sales went up after we instituted our new advertising campaign? Our success is obvious." (It is not enough simply to deny causation; one must suggest other possible alternative explanations for the event in question.)
Slippery Slope	Assuming, unjustifiably, that a proposed step will set off an undesirable and uncontrollable chain of events	Today it is abortion, but tomorrow it will be the mentally ill, and then the infirm and the aged—or anyone else considered undesirable.
Irrelevant Thesis	Seeking, perhaps succeeding, to prove a conclusion not at issue (It can take the form of either (1) attacking someone else's claim, irrelevantly, or (2) defending a claim of one's own, irrelevantly.)	(1) "The advocates of conservation contend that if we adopt their principles we will be better off than if we did not adopt them. They are mistaken, for it is easy to show that conservation will not produce an Eden on earth." (2) "I fail to see why hunting should be considered cruel when it gives tremendous pleasure to many people and employment to even more."

fallacy of sweeping generalization, the error lies in assuming that what is true under certain conditions must be true under all conditions. In the fallacy of hasty generalization, the error lies in assuming that the evidence on which the argument is based is sufficient to warrant its conclusion, when in fact such evidence is either unrepresentative or insufficient. Finally, the fallacy of bifurcation assumes falsely that the alternatives presented in the argument are the only alternatives available, when other alternatives do exist.

1. THE FALLACY OF SWEEPING GENERALIZATION

The *fallacy of sweeping generalization* is committed when a general rule is applied to a specific case to which the rule is not applicable because of special features of that case. Consider this example:

>   a)   Everyone has a right to his or her own property. Therefore, even though Jones had been declared insane, you had no right to take away his weapon.

The first premise in this argument is a general principle that is widely accepted. It does not apply, however, in the specific case in which a person has lost his or her reason—especially when the piece of property is a weapon.

The source of this fallacy's persuasive power is that it resembles valid arguments in which individual cases do fall under a general rule. The point to remember is that a generalization is designed to apply only to individual cases which properly fall under it. It is not designed to apply to all individual cases.

It is certainly valid to argue that, since all men are mortal, Socrates is mortal. But it would be incorrect to argue:

>   b)   Since horseback riding is healthful exercise, Harry Brown ought to do more of it because it will be good for his heart condition.

What is good for someone in normal health does not apply where special health problems exist. The fallacy of sweeping generalization is also referred to as the *fallacy of accident,* to emphasize this

"Calm down, Edna . . . Yes, it's some giant, hideous insect . . . but it could be some giant, hideous insect in need of help."

"The Far Side" cartoon by Gary Larson is reprinted courtesy of Chronicle Features, San Francisco.

Edna's friend is trying to prevent Edna from making the fallacy of sweeping generalization. What is true of most hideous insects might not be true of this particular one, at this particular point in time.

characteristic of the irregularity of particular cases to which generalizations do not apply.

Although the examples of sweeping generalization noted thus far are relatively clear-cut, this fallacy can be difficult to untangle when the ideas involved are more complex. Consider this argument:

c) I believe in the golden rule as an inherent duty to do unto others as I would have them do unto me. If I were puzzled by a question on an examination, I would want my neighbor to help me. So it is my duty to help the person sitting next to me who has asked me to give her the answer to a question on this exam.

A critical fact has been overlooked here. The purpose of an examination is to find out how much each person knows, and that purpose would be defeated if people helped each other. If that is so, the golden rule does not apply in these circumstances.

An argument of the kind we are examining has two parts to it—a rule and a case. If the argument is invalid, it is because the case to which the rule is being applied is exceptional and therefore does not fall under the given rule. To expose such a fallacy, therefore, all one needs to do is to isolate the rule and show that if understood properly it cannot be applied to the case in question. Reviewing our examples, we might say of them: *(a)* yes, everyone has a right to his or her own property—if that person has not gone insane; *(b)* yes, horseback riding is healthful—if one's state of health permits it; *(c)* yes, we ought to help each other—but not when an examination is involved.

Like some of the other fallacies studied thus far, this one too can lend itself to humorous effects. A well-known example has come down to us from Boccaccio's *Decameron,* a witty collection of tales from fourteenth-century Italy.

Somewhat abbreviated, the incident reads:

> d) A servant who was roasting a stork for his master was prevailed upon by his sweetheart to cut off a leg for her to eat. When the bird came upon the table the master desired to know what was become of the other leg. The man answered that "the stork never had but one leg." The master, very angry, but determined to strike the servant dumb before he punished him, took him the next day into the fields, where they saw storks standing each on one leg, as storks do. The servant turned triumphantly to his master, upon which the latter shouted, and the birds put down their other leg and flew away. "Ah, sir," said the servant, "but you did not shout to the stork at dinner yesterday; if you had done so, he would have showed his other leg too." (Sixth day, fourth tale)

The rule in this case can be phrased as: Storks who customarily stand on one leg will put down their other leg when shouted at. But it does not apply in cases of storks who are either one-legged to begin with or roasted.

It should be borne in mind that, although generalizations can be abused, they remain extremely useful in logic. Generalizations allow us to infer universal rules with a reasonable degree of certainty where we could not possibly inspect each and every case. Indeed, to become so technical or precise that we avoid a generalization because we can conceive of an isolated case that might be an exception is to engage in quibbling or pettifogging. Lawyers with a good knowledge of the law and a fine eye for detail will sometimes succeed in having charges dropped against a client on the basis of a technicality—of what in other contexts might be considered pettifogging. But such things are matters of degree. To fail to recognize mitigating circumstances when judging a person is to serve only the letter of the law, not its spirit. It is, again, to apply the law mechanically while overlooking important facts.

2. THE FALLACY OF HASTY GENERALIZATION

The *fallacy of hasty generalization* is precisely the reverse of sweeping generalization. In hasty generalization an isolated or exceptional case is used as the basis for a general conclusion which is unwarranted. Consider two examples:

a)   I had a bad time with my former husband. From that experience I've learned that all men are no good.
b)   I know one union representative and she's a terrible person. I wouldn't trust any of them.

Both arguments are invalid because they assume that what is relatively true under certain conditions is true under all conditions. At most the "evidence" presented here (if one may call it that) warrants only a particular, not a general, conclusion.

Unlike the fallacy of sweeping generalization, which results when a generalization is misapplied, the fallacy of hasty generalization results when a particular case is misused. Here too, it must be remembered that it is usually impossible—and unnecessary—to examine all possible cases. Unless a sufficient number are tested, however, the conclusion is likely to be too hasty and therefore unreliable.

In some instances the fallacy of hasty generalization results from resting a conclusion upon cases which are exceptional and therefore unrepresentative. Thus we would be guilty of hasty generalization if we were to say, "He speaks so beautifully that anyone can see he must have studied acting," or "This must be good medicine because it tastes awful." In both situations, the particular cases are related to the general rule only in an unessential way. There is no basis for assuming a connection.

The Sherlock Holmes stories by Arthur Conan Doyle abound with illustrations of complicated reasoning by the master detec-

"Here's the last entry in Carlson's journal: 'Having won their confidence, tomorrow I shall test the humor of these giant but gentle primates with a simple joy-buzzer handshake.'"

"The Far Side" cartoon by Gary Larson is reprinted courtesy of Chronicle Features, San Francisco.

A hasty generalization. What is amusing to Professor Carlson, and perhaps a few of his fellow primates, might not be amusing to all *primates.*

tive. Only seconds after being introduced to Dr. Watson, Holmes brilliantly "concludes" that Watson must have been in Afghanistan.

> c) Here is a gentleman of a medical type, but with the air of a military man. Clearly an army doctor, then. He has just come from the tropics, for his face is dark, and that is not the natural tint of his skin, for his wrists are fair. He has undergone hardship and sickness, as his haggard face says clearly. His left arm has been injured. He holds it in a stiff and unnatural manner. Where in the tropics could an English army doctor have seen much hardship and got his arm wounded? Clearly in Afghanistan. ("A Study in Scarlet," pt. 1, ch. 2)

Obviously, Watson could have had an air of a military man without ever having been in the army; he could have got his face tanned without leaving the British Isles; and so forth. Holmes is guilty of a hasty generalization founded on insufficient evidence.

A variant of hasty generalization occurs when only that evidence which supports an argument is brought into the argument, while evidence to the contrary is ignored.

> d) Under the capitalist system many people are poor, and there is waste of labor and materials, cutthroat competition, glorification of the acquisitive instinct, depression on the one hand and inflation on the other. All this proves that the system is thoroughly rotten and ought to be discarded.

Even assuming that all the evidence in this argument is correct, it would not constitute a sufficient basis for the drastic action recommended. To try to correct the abuses by abolishing the system is like throwing the baby out with the bath water. Nor has any attention been given to advantages of the system which may mitigate its disadvantages.

The thinking behind argument *d* is a form of *rationalization,* the giving of seemingly sound reasons to justify a belief held on other, often less respectable grounds. Rationalization is used by those who have vested interests to protect. Rather than examine all the evidence, they select only that which favors their interests.

At the same time, they try to convey the impression that all the evidence has been examined and that all of it is on their side. Certain politicians thus view all their own achievements with pride and all those of the opposition with alarm. Or they appeal for votes on the ground that they will reduce taxes but do not call attention to reductions in government services that will result from such cuts. Loan companies stress how easy loans are to secure and how small monthly payments are but do not mention the relatively high interest rates or how difficult some people find it to repay loans. Matters are rarely so simple as rationalization makes them seem.

One place such practices might seem permissible is in a court of law, where it is expected that each side will present its own case in the most favorable light. Even here, however, the principle of cross-examination has evolved in order to ensure that judge and jury attend to all of the evidence and hear the whole truth.

Reasons of some magnitude can be found against virtually any argument, yet we do not expect arguments to include reasons against as well as for their conclusions. The fallacy of hasty generalization is committed when we fail to take account of facts that are of such significance that their consideration would require a rational person to reject our conclusions. We risk committing this fallacy when we attend only to evidence which supports our case, failing to consider evidence to the contrary.

3. THE FALLACY OF BIFURCATION

The *fallacy of bifurcation** is an argument which presumes that a distinction or classification is exclusive and exhaustive, when other alternatives exist. Bifurcation is intimately bound up with

---

Bifurcation comes from the Latin *bifurcus,* meaning "two-pronged." (*Bi* is the Latin prefix for "two," and *furca* means a "fork" or "branch.")

---

confusion over the words *either/or.* "Either we must have war against Russia before she has the atom bomb," Bertrand Russell argued in 1948, "or we will have to lie down and let them govern us." In the debate some chanted, "Better dead than red!" while others replied, "Better red than dead"—both sides overlooking a third position: "Better pink than extinct."

This fallacy presents *contraries* as if they were *contradictories*. Two statements are said to be contraries when it is impossible for both to be true but possible for both to be false. If we say that Jane may be rich or she may be poor, for example, we mean that she cannot be both at the same time but that she may be neither. Two statements are said to be contradictories, on the other hand, when it is impossible for both to be true and also impossible for both to be false. Either that man is alive or he is dead. Either today is your birthday or it isn't. If one contradictory is true, the other must be false, and vice versa. The fallacy of bifurcation arises when an either/or statement that actually contains two contraries is instead put forward as containing two contradictories.

Because our language is full of opposites, the tendency to bifurcate is common. We are prone to people the world with the "haves" and "have-nots," the "good" and the "bad," the "normal" and the "abnormal"—forgetting that between these extremes lie numerous gradations, any of which could serve as further alternatives to an *either/or* polarity.

Bifurcation is typical of, for example, the busy administrator who cries, "Spare me the details! Is the report good or bad?"—as if it could not be a bit of both. It was typical of those who spoke out strongly against everything Russian during the cold war.

> a)   Let me give this solemn warning. There can be only one capital, Washington or Moscow. There can be one flag, the Stars and Stripes or the godless hammer and sickle. There can be one national anthem, the "Star-Spangled Banner" or the Communist *"Internationale."*

And in a less profound but equally exasperating vein, it is typical of anyone who counters a complaint in this manner: "So you think the soup is too cold, do you? Well, I suppose you'd like it scalding hot!" What is objectionable about such arguments from a logical point of view is that they force us into a position of choosing between two alternatives when in fact our choices are not so limited.

Thinking in extremes can be appealing, unfortunately, for it requires less mental energy than exploring all aspects of a prob-

Your Choice.

The argument in this advertisement for a speed-reading course is: either you have not taken the course and are weighed down by work, or you have taken the course and are now free from work and worry. If it were only that simple! Fortunately, there are other alternatives.

lem. Advertisers often try to cut off our critical thinking about a product by channeling our view of it into an either/or polarity that suits their aim.

b) If you know about BMW, you either own one or you want one.

Sometimes it is our prejudices that blind us to additional alternatives, but sometimes it is simply our desire for a simple solution to a complex problem. It is tempting to suppose, for example, that our foreign policy could be based on the formula that other nations are either our friends or our enemies. Then we would not have to struggle with the fine shades of differences that actually characterize them.

In some instances of the fallacy of bifurcation, two terms are made to seem contradictory when they are not even contrary.

c) We must choose between safety and freedom. And it is in the nature of good Americans to take the risk of freedom.

Safety and *freedom* are not necessarily incompatible. By implying that they are, this argument commits a fallacy, for it invites us, again, to overlook facts that bear directly on the argument.

EXERCISES

Identify the fallacy of presumption—sweeping generalization, hasty generalization, or bifurcation—that is committed in or that could result from each of the following. Explain the errors committed in each case.

* 1. Shakespeare cannot have been a great writer, for he did not even make up his own plots.

B 2. There are two kinds of people in this world: the "haves" and the "have-nots."

* 3. If we're going to buy a car, we have to buy either a good one or a cheap one. We can't afford a good one, and we don't want a cheap one; so we'll just have to do without a car.

* 4. Doctors are all alike. They really don't know any more than you or I do. This is the third case of faulty diagnosis I've heard of in the last month.

SG 5. Since it is right to speak the truth, it is therefore right to tell our friends what we think of them.

* 6. Only the rich and the poor need concern the government: the rich because they try to influence legislators through the power of their wealth; the poor because they have no wealth to care for themselves.

HB 7. If you work hard, you will get good grades. (Overheard)

* 8. At my interview, a Social Security Administration officer tried to explain to me the earnings limitations rules for monthly income but gave up after his third attempt, declaring he really didn't understand it himself. How does the Social Security Administration expect people like me to understand its rules, when its own workers don't know what's going on.

* 9. Either she knew everything that was going on, in which case she's a liar, or, alternatively, she's a fool and knew nothing.

10. People either ask for Beefeater or they ask for gin. (Advertisement)

*11. Is it lawful to do good on the Sabbath days, or to do evil? (Jesus of Nazareth [Luke 6:9]) (when accused of violating the prohibition against working on the Sabbath by healing a man with a withered hand)

12. They just don't care about traffic law enforcement in this town, for they let ambulances go at any speed they like and let them run red lights too. (Overheard)

13. Death should be held of no account, for it brings but two alternatives: either it utterly annihilates the person and his soul, or it transports the spirit or soul to someplace where it will live forever. What then should a good man fear if death will bring only nothingness or eternal life. (Cicero, ancient Roman statesman and philosopher)

14. DEAR ABBY:

Thanks for advising JUST PLAIN JEALOUS to trust her husband on those business trips with a female co-worker.

I'm an airline stewardess who's engaged to be married soon, and if my fiancé didn't trust me, I'd take it as an insult.

Everyone seems to think stewardesses and pilots fool around a lot, but it's not true.

What would an intelligent, good-looking girl in her twenties want with a balding, middle-aged, burned-out guy who's old enough to be her father? Besides that, most pilots are either broke from paying alimony and child support, or they've got a couple of kids to send to college and a house in the suburbs that's not paid for.

No thanks!

*15. Let me warn you that you will find in the world a certain few scoffers who will laugh at you and attempt to do you injury. They will tell you that John D. Rockefeller was a thief and that Henry Ford and other great men are also thieves. Do not believe them. The story of Rockefeller and of Ford is the story of every great American, and you should strive to make it your story. Like them, you were born poor and on a farm. Like them, by honesty and industry, you cannot fail to succeed. (Nathaniel West, *A Cool Million*)

16. The Constitution guarantees freedom of speech. Therefore, if a person believes that the only way to achieve cer-

tain reforms is through rioting, he or she should have the freedom to incite to riot.

EVADING THE FACTS

In this second category of the fallacies of presumption, the error lies not in overlooking facts as in the first category, but in seeming to deal with all relevant facts without actually doing so. Such arguments deceive by inviting us to presume that the facts are as they have been stated in the argument, when the facts are quite otherwise.

Four fallacies commit this error. The fallacy of begging the question tries to settle a question by simply reasserting it. Question-begging epithets avoid a reasonable conclusion by prejudging the facts. Complex questions evade the facts by arguing a question different from the one at issue. And, finally, special pleading invites us to view the argument from a biased position.

4. THE FALLACY OF BEGGING THE QUESTION

The *fallacy of begging the question* is committed when, instead of offering proof for its conclusion, an argument simply reasserts the conclusion in another form. Such arguments invite us to assume that something has been confirmed when in fact it has only been affirmed or reaffirmed. If the argument is a simple one, few will be taken in by it.

    a)   The belief in God is universal because everybody believes in God.

Since *universal* means "something applicable to *everybody*," this argument simply reasserts what is at issue without proving it. In effect, such an argument lacks premises and is therefore not an argument at all, for all we have done is to repeat the very point with which we began. The repetition of a conclusion in another form should never be mistaken for proof of that conclusion.

Mayor Richard Daley of Chicago employed this fallacy for humor—and to evade the facts, perhaps—when a reporter asked him why Senator Hubert Humphrey had failed to carry Illinois in an election. Humphrey lost the state, replied Daley, because "he didn't get enough votes." The question really was: Why did he not get enough votes? Less intentionally humorous was President Calvin Coolidge's unfortunate remark that "when large numbers of people are out of work, unemployment results." The "arguments" here—merely explaining, as they do, their own terms—are unsound because to explain one's terms is not to establish their truth or validity. As in Molière's famous play where a physician explains the soporific effect of opium by citing its sleep-producing properties, it is a matter of explaining a thing by the very thing that needs explaining. And this points up the essential circularity of such reasoning. Such "explanations," we might say, are not even explanations. To say that Tom committed suicide because he had a death wish, or that bodies fall because they have a downward tendency, is to reveal nothing other than that Tom committed suicide or that bodies fall.

Flagrant examples of this fallacy escape easy detection if the statements involved are somewhat drawn out, in which case our memories may fail to spot the repetition. Consider this example:

b)   Free trade will be good for this country. The reason is patently clear. Isn't it obvious that unrestricted commercial relations will bestow on all sections of this nation the benefits which result when there is an unimpeded flow of goods between countries?

Because "unrestricted commercial relations" is just a more verbose way of saying "free trade," and since the wordy conclusion of the argument is merely another way of saying "good for this country," the argument says, in effect, that free trade will be good for the country because free trade will be good for the country. Arguments that beg the question are *circular arguments.* They make use of the capacity of our language to say a thing in many different ways, ending where they began and beginning where they end. They are like the proverbial three morons, each of whom tied his horse to another's horse, thinking that he had in this way secured his own horse. Naturally, all three horses wandered away because they were anchored to nothing but each other.

The circle that is involved in such arguments can be appreciated in this scene, which takes place in a loan company office:

c) MANAGER: And how does our loan company know that you are reliable and honest, Mr. Smith?

SMITH: Well, I think Mr. Jones will give me a good reference.

MANAGER: Good. But can my company trust the word of Mr. Jones?

SMITH: Indeed it can, Sir! I vouch for the word of Mr. Jones myself!

If Jones is to vouch for Smith and Smith for Jones, we are no further ahead than we were at the outset.

Circular arguments reason that A is so because of B. But B turns out to be true only if A is true. The question, Is A true? remains unanswered. The question is begged. Here is a familiar example:

d) God exists!
How do you know?
The Bible says so.
How do I know that what the Bible says is true?
Because the Bible is the word of God!

Such arguments are sometimes called *vicious*—as in the familiar phrase, *vicious circle*—to indicate that wherever we try to break through them we are forced back again.

e) People can't help doing what they do.
Why not?
Because they always follow the strongest motive.
But what is the strongest motive?
It is, of course, the one that people follow.

We have seen that a sound argument provides reasons for a conclusion. Because of the very fact that the conclusion to be established is somehow in doubt, an attempt is made to support it by premises that are more certain. But if the supporting premises merely repeat what is stated in the conclusion, as in all cases of begging the question, the argument contains no premises and is therefore fallacious.

One common form of the fallacy of begging the question is the use of an unfounded generalization to support a conclusion that would fall under the generalization if it were true.

> f) Government ownership of public utilities is a dangerous doctrine, because it is socialistic.

If the larger thesis (that socialism is dangerous) were granted, then the particular principle being argued here—that government ownership of public utilities is dangerous—would follow. Our opponent may, however, regard our assumption about socialism as the very thing in question.

Consider this example:

> g) His cruelty may be inferred from his cowardice, for all cowards are cruel.

Here again a principle wider than the conclusion itself is used as proof for the conclusion. Because the truth of that principle may be questioned, however, to use it as support for the conclusion is to beg the question. If the conclusion needs proving (being merely an instance of the wider principle), surely the wider principle needs proving all the more so.

The fallacy of begging the question can also assume the opposite form. Whereas argument *g* used a suspect generalization as proof for a particular case, it can also occur that a suspect particular is used as proof for a generalization.

> h) I don't need to hear the testimony in this case, because I already understand it. It's another case of a young person killing an older person, isn't it? Well, I know that all such cases are the result of childhood environment. This case has just got to be the result of childhood environment, because it even strengthens my belief that all murders of old people by younger people are the result of the murderer's childhood experiences.

Here the particular case of one crime allegedly committed as the result of childhood environment is used as support for the universal assumption that all such crimes are rooted in childhood environment. But since it is not at all obvious that this generalization is itself true, it needs further support. It cannot get this support from the particular case that it itself is called in to support, however, without making a complete circle. We have, then, a particular supported by a doubtful universal whose sole support is this very particular. As in the example of the loan company in argument *c,* we can say that if the particular could support the universal, we should not need the universal to begin with. Since we obviously do need it, it cannot rest on that very particular which rests on it.

To beg a question in the form we have just examined is to try to establish an uncertain statement on the basis of one that may seem even more uncertain. To argue that a particular crime is the result of childhood environment because all such crimes are rooted in childhood experiences would be logically objectionable even if the argument did not make a complete circle. For the generalization upon which the particular is based may be one whose truth is not accepted by one's opponents. They will therefore regard our argument as begging the question. By contrast, we might maintain that our reasoning is perfectly sound because, after all, we are merely drawing out implications contained within the given premises. For if it is true that all such crimes are rooted in childhood, then it follows that this crime too must be similarly rooted. If to argue so is fallacious because circular, then perhaps all arguments are fallacious.

There is some point to this observation. All arguments, attempting as they do to draw out implications contained in their premises, are in a sense circular. There are two respects, however, in which sound arguments differ from those that we condemn because of their circularity.

First, a sound argument contains premises that assert information not contained in the conclusion. That is, the premises provide evidence for the conclusion rather than simply restating assertions made in the conclusion, as circular arguments do. Second, a sound argument may contain obviously true premises rather than premises which are open to question. An obviously true premise, for example, would be that someone cannot be in two places at the same time; anyone using this premise in an argument would have to be granted it without having to demonstrate it. Circular arguments, on the other hand, are attempts to persuade us of a doubtful conclusion by presenting us with evidence that is equally suspicious.

It is useful to recall that the fallacy of begging the question, like all fallacies, is most easily detected when we examine the argument from the standpoint of what we are rationally obliged to grant within the structure of a given argument. If we are told, for example, that suicide is a crime because it is a crime to commit murder, we are not obligated to accept the underlying assumption that taking one's life is murder.

5. THE FALLACY OF QUESTION-BEGGING EPITHETS

In the *fallacy of question-begging epithets,* the error lies in the use of slanted language that reaffirms what we wish to prove but have not yet proved. An *epithet* is a descriptive word or phrase used to characterize a person, a thing, or an idea. A favorite device of poets, the epithet has little place in logic. We have seen that to beg the question is to assume the point in dispute. We shall now see that it is possible to do this with a single word or phrase, one which suggests that a point at issue has already been settled when in fact it is still in question. Since many of our words possess both a descriptive and an evaluative dimension, the possibilities for question-begging epithets are endless.

Consider the implications underlying this statement:

> a) This criminal is charged with the most vicious crime known to humanity.

One might rather have said, "This person is charged with homicide," keeping to the facts rather than attaching prejudicial labels that support a particular opinion about the matter.

The term *stealing,* for example, does not merely describe but also makes a judgment about an act, for the term suggests that the action is wrong. In the case, say, of a man "stealing" food in order to prevent his children from starving to death, the use of such a word would beg the question, since in light of the circumstances the man's action would probably be considered appropriate.

The fallacy of question-begging epithets is so widespread that it has come to be known by a variety of labels: *mud slinging, name calling,* use of *loaded words* and *controversial phrases, verbal suggestion,* and *emotive language.* Question-begging epithets are objectionable because their object is to arouse our passions and prejudices through the use of emotionally charged language. By overstatement, ridicule, flattery, abuse, and the like, they seek to evade the facts, leading us to believe that the terms so used properly describe the people or events in question when this may not be so at all.

The following are some question-begging epithets that render their arguments fallacious:

b) When one reflects on how this country's unemployment insurance program has encouraged the indolence of skid-row bums and Cadillac-driving welfare cheats, it becomes obvious that this legalized highway robbery must be revoked.

c) The council member's shocking proposal is calculated to subvert the just aspirations of hard-working citizens.

d) No right-thinking American could support this measure, a cunning plot hatched in the smoke-filled rooms where bossism rules.

The question-begging epithets used in these arguments are logically objectionable because they assume attitudes of approval or disapproval without providing evidence that such attitudes are justified. To call someone a "welfare cheat" or a "hard-working citizen" is not to establish that the name fits him or her; nor does calling a certain measure a "cunning plot" or "shocking" make it so.

We can observe, also, that one may beg the question not only with *dyslogistic* (uncomplimentary) epithets but with *eulogistic* (complimentary) ones as well. When we refer to a certain historical event as the "Reformation," we may be guilty in the eyes of some historians of assuming the point in dispute, since the word *reform* means not simply a change but a change for the better. Similarly, "right-thinking American" is a flattering epithet for "American."

One of the most famous instances of the dyslogistic form of this fallacy in recent years was former Vice President Spiro Agnew's remark about dissenters. In a speech delivered in New Orleans, Agnew charged:

e) A spirit of national masochism prevails, encouraged by an effete corps of impudent snobs who characterize themselves as intellectuals.

Cartoonist Al Capp improvised an unkind epithet for students walking out in protest from a lecture he was delivering in Hartford: "Hey! Don't go. I need an animal act."

General semanticist S. I. Hayakawa provides a more humorous example of the confusion between fact and opinion that underlies the fallacy of question-begging epithet. He sketches this scene from a cross-examination during trial:

f) WITNESS: That dirty double-crosser Jacobs ratted on me.
DEFENSE ATTORNEY: Your honor, I object.
JUDGE: Objection sustained. (Witness's remark is stricken from the record.) Now, try to tell the court exactly what happened.
WITNESS: He double-crossed me, the dirty, lying rat!
DEFENSE ATTORNEY: Your honor, I object!
JUDGE: Objection sustained. (Witness's remark is again stricken from the record.) Will the witness try to stick to the facts.
WITNESS: But I'm telling you the facts, your honor. He did double-cross me.

"This can continue indefinitely," Hayakawa points out, "unless the cross-examiner exercises some ingenuity in order to get at the facts behind the judgment. To the witness it is a 'fact' that he was 'double-crossed.' Often patient questioning is required before the factual bases of the judgment are revealed" (*Language in Thought and Action,* 4th ed. New York: Harcourt Brace Jovanovich, 1978, pp. 37–38). A question-begging epithet is an indication that relevant facts are being evaded, for it invites us to join with the formulator of the argument in prejudging the issues. A sound argument might convince us that the epithet is appropriate, but such slanted language has no place in the argument itself.

6. THE FALLACY OF COMPLEX QUESTION

The *fallacy of complex question* is the interrogative form of the fallacy of begging the question. Like the latter, it begs the question

"If elected, would you try to fool some of the people all of the time, all of the people some of the time or go for the big one: all of the people all of the time?"

Reprinted with permission from *TV Guide* ® Magazine and Henry R. Martin. Copyright © 1980 by News America Publications Inc., Radnor, Pennsylvania.

Before rushing to answer a complex question, it is best to question the question: "Why do you think I am trying to fool people?"

by assuming the conclusion at issue. Complex question accomplishes this by leading one to believe that a particular answer to a prior question has been answered in a certain way when this may not be the case. This fallacy goes by many names, including *loaded question, trick question, leading question, fallacy of the false question,* and *fallacy of many questions.*

It is told of King Charles II of England that he once asked members of the Royal Society to determine for him why it is that if you place a dead fish in a bowl of water it makes the water overflow, while a live one does not. Some of the members thought about this a very long time and offered ingenious but unconvincing explanations. Finally, one of them decided to test the question. He discovered, of course, that it did not make a bit of difference whether one placed a dead fish or a live one in the bowl of water.

Whether the story is true or not, it holds an important lesson. Before rushing to answer a question, it is best to question the question. For every question necessarily brings with it a set of assumptions which determine the lines along which it is to be answered, and sometimes those assumptions may render the argument fallacious. For example:

a)   Have you stopped beating your wife?
b)   Did John ever give up his bad habits?
c)   Are you still a heavy drinker?

In each of these questions there lies an assumed answer to a previous question. Did John have bad habits? is the unasked question whose answer is assumed in question *b*. We need to withhold any answer to question *b* until this prior question has been resolved. In some instances of this fallacy, considerable struggle may be necessary in order to liberate ourselves from the misleading influence of a complex question.

The serious consequences of complex questions can be appreciated by considering these trick questions, which would be out of order in a court of law:

d)   What did you use to wipe your fingerprints from the gun?
e)   How long had you contemplated this robbery before you carried it out?

Both questions rest upon an answer to a prior question that has been neither asked nor answered in a way required by the complex question.

Such questions are fallacious, also, because they assume that one and the same answer must apply to both the unasked and the asked question.

f)   Isn't Jane an unthinking radical?

This question demands an unqualified yes or no, yet if we divide the question into its two parts we may well wish to reply: yes, she is a radical, but she is not an unthinking one.

A complex question may often appear in combination with a question-begging epithet.

g)   What are your views on the token effort made by the government to deal with this monstrous oil spill?
h)   Was it through stupidity or through deliberate dishonesty that

the administration hopelessly botched our relations with Iran?

Before they can be dealt with logically, complex questions must be divided, not only into prior and subsequent questions but also into their descriptive and evaluative elements. In argument *h,* for example, we would need to establish not only whether the administration had been a cause of worsened relations with Iran but, if it had, whether it did so through actions that could be labeled *stupid, dishonest,* or *hopeless.*

Complex questions invite us to evade the facts by their very complexity, amid which we may lose sight of the point at issue. In this vein, the following report appeared in the pages of the *Times* (London) (February 22, 1971):

> i)   Lord Boothby asked the government whether the Minister of Aviation still contemplated raising the landing fees at London Airport on April 1 to the point which would make it not only the worst but by far the most expensive airport in the world.

This argument raises not only the question whether raising the landing fees would make London Airport the most expensive in the world but also the issue of whether it is already the worst airport in the world. These two questions may impinge upon one another, but they should be asked separately.

Complex question seems a favorite with advertisers and salespeople. An ad for sleeping tablets asks: "Does the sleeping tablet you're now taking start to work in 21 seconds?" One for office security systems inquires: "Can you afford to continue to run the risk of having your office accessible to prowlers?" A sales representative asks whether you'll be paying by check, cash, or credit card—long before you've decided whether to buy at all. Or:

> j)   Aren't your good friends worth your best Bourbon?
> k)   When should you buy your first Cadillac?

By assuming the positive merits of the products, these advertisements narrow their arguments to the less weighty question of when we will avail ourselves of the products—the advantages of which have never been established.

Complex questions can take the form of asking for an explanation for "facts" which are either untrue or not yet established.

l) Why is a ton of lead heavier than a ton of feathers?
m) What is the explanation for mental telepathy?

By focusing our attention on explaining the "facts," such questions divert our attention from the fallaciousness of the questions themselves.

The study of complex question teaches that questions should be asked one at a time and that no attempt should be made to answer a question until the one on which it depends has been settled. This principle is acknowledged in legal and parliamentary procedure, where the rules allow for a motion "to divide the question." This procedure provides for the fact that questions may be so complex that they should be considered only when separated.

7. THE FALLACY OF SPECIAL PLEADING

To commit the *fallacy of special pleading* is to apply a double standard: one for ourselves (because we are special) and another (a stricter one) for everyone else. Bertrand Russell once illustrated this all-too-human characteristic by showing how we "conjugate" certain words—*firmness* for example.

a) I am firm; you are stubborn; he is pigheaded.

When we engage in special pleading, we favor ourselves and are prejudiced against others. As in the case of question-begging epithets, we imply—and hope others will believe—that our labeling correctly describes reality when in fact it merely reflects our prejudice.

Like all fallacies, this one can be exploited for its humorous effects. When asked her opinion of on-stage nudity, the actor Shelley Winters (then aged forty-six) replied:

b) I think it is disgusting, shameful, and damaging to all things American. But if I were twenty-two with a great body, it would be artistic, tasteful, patriotic, and a progressive, religious experience.

To engage in special pleading is to be partial and inconsistent. It is to regard one's own situation as privileged while failing to apply to others the standard we set for ourselves (or, conversely, to fail to apply to ourselves those standards we apply to others). It is to speak, for example, of the *heroism* of our troops, their *devotion* and *self-sacrifice* in battle, while describing the enemy as *savage* and *fanatical,* as in the following remark from an editorial:

> c) The ruthless tactics of the enemy, his fanatical, suicidal attacks, have been foiled by the stern measures of our commanders and the devoted self-sacrifice of our troops.

Do *self-sacrifice* and *suicide* describe different events? Obviously they do not, but such use of language hopes to persuade others that the difference in labels reflects a difference in the quality of the events in question.

The existence in our language of a large body of paired terms such as *heroism/savagery* makes it all too easy to engage in special pleading without being fully aware that we are doing so. In the list below, both terms in each set have the same descriptive content but each differs from the other in the attitude that it invites or leads us to infer is appropriate.

work/toil
sportsman/playboy
police officer/cop
group/gang
active/noisy
talk/jabber
plan/scheme

Through special pleading, bad habits—parsimony and lack of hygiene—are turned into a good one, thrift.

enterprising/opportunistic
payment/dole
execution/slaughter
colorful/loud
overweight/fat
earthy/gross
cautious/negative
reserved/secretive

Indeed, much of what we say can be seen as special pleading, as we apply one standard here, another there.

Efforts by feminist groups have attempted to focus attention on double standards. Women's objections to being called "girls" by men is a case in point. Where such consciousness raising is effective, it tends to bring about changes in language usage, as in recent coinages such as *chairperson* and *spokesperson.* Similar consciousness raising is useful for students of logic who wish to purge their arguments of language and attitudes implying a double standard at variance with the facts.

EXERCISES

Identify the fallacy—begging the question, question-begging epithet, complex question, or special pleading—that is committed in or that could result from each of the following. Explain the error committed in each case.

* 17.   Haste makes waste, because hurried activity is always careless activity.

* 18.   "We're worried about your son, Mr. Norris," says the sixth-grade teacher. "He seems lazy. He persuades young Stewart to do all his work." "Lazy!" exclaims Norris, "That's executive ability."

* 19.   Which one of you left the door open?

* 20.   I'm surprised at you. A person of your culture and upbringing—defending these hoodlums!

* 21.   The world was not created by God, for matter has always existed and therefore the world must have always existed.

* 22.   The new bell in the chapel is louder than the old one. The old one didn't make nearly as much noise.

* 23.   Why are the students in this class more intelligent than others in the university?

  24.   I will not commit this act because it is unjust. I know it is unjust because my conscience tells me so; and my conscience tells me so because the act is wrong.

* 25.   On November 5, three of the accused met at the house of the fourth defendant, Smith. There, behind locked doors and heavily curtained windows, these four conspirators began to hatch their dastardly plot.

  26.   Everybody knows that people who sell life insurance are doorbell-ringing, back-slapping, neighbor-nagging pests who spend their days trying to sell all the insurance they can to anybody who'll talk to them, if anybody will talk to them, and they're probably in the business in the first place because they flunked out of college and couldn't find a better job. Everybody knows that, right?

* 27.   As one matures in life, I feel one ought to increasingly learn to lean on Jesus Christ and not on other props. Personally, I'm committed to get along without chemicals as much as I can. I seldom take aspirin, feeling that most headaches come because of tension and that it's better for me to learn to adjust my life so tension doesn't play such a major role. I don't want to rely on chemicals for my well-being.

         I regard marijuana as a very obvious crutch. Unlike alcohol, no one uses it for the taste. The only reason to smoke dope is to alter your consciousness, to escape a little bit from reality. But I don't want to escape from reality. I think that the argument against marijuana here exactly parallels what Paul wrote in Ephesians 5:18: "Don't drink too much wine . . . be filled instead with the Holy Spirit." Rather than dealing with life by smoking dope or getting drunk, I'd rather find God's strength (and excitement) to face problems and overcome them. (Religious circular)

  28.   Who made God? (Child's question)

* 29.   Isn't it finally time you bought the best? (Advertisement for a stereo system)

* 30.   Comedian W. C. Fields said he knew a sure cure for insomnia—a good rest.

  31.   Health-food-store hucksters who attempt to diagnose symptoms of deadly diseases and prescribe worthless "cures" are recklessly endangering people's lives.

32. Your noble son is mad:
    Mad call I it, for to define true madness,
    What is't but to be nothing else but mad?
    (Polonius in Shakespeare's *Hamlet*)

33. This measure ought to be deplored by all right-thinking people.

34. This is the same cynical, rotten, misleading bull-roar that the oil companies have been handing us all along. Why should we continue to listen to these selfish bastions of entrenched interest and misbegotten wealth? How can we be so shortsighted? Critical oil shortage, my asthma! Miasma is what it is—a poisonous foul-smelling vapor of smog and oil company propaganda.

35. Obviously, we didn't have enough money in the bank. (Charlie Theokas, general manager of the New Jersey Nets, on the paychecks of three players that bounced)

36. Let's stock up before the hoarders get here.

37. I can't stand my new house. There's almost no privacy. I see people wearing my clothes when I never gave them permission. Not only do things disappear from the closet, but they even go into my drawers. There was one time when I didn't even know that one of my blouses was missing until I ran across it in someone's drawer.

DISTORTING THE FACTS

In this third and final part of our chapter on the fallacies of presumption, we shall consider fallacies that, rather than overlooking or evading relevant facts, actually distort such facts. In the fallacy of false analogy, certain cases are made to appear more similar than they really are. The fallacy of false cause makes it appear that two events are causally connected in a way they are not. The fallacy of slippery slope makes it seem that a certain position will set off an undesirable chain reaction when this may not be so at all. And the fallacy of irrelevant thesis distorts by concentrating on an issue which is actually irrelevant to the argument.

8. THE FALLACY OF FALSE ANALOGY

Few techniques of reasoning are so potentially useful—or so potentially dangerous—as analogy. When we reason by analogy we attempt to advance our position by likening an obscure or difficult set of facts to one that is already known and understood and to which it bears a significant resemblance. The *fallacy of false analogy* arises when the comparison is an erroneous one that distorts the facts in the case being argued.

Drawing attention to likenesses can be extremely useful so long as the two things being compared resemble each other in important respects and differ only in trifling ways. If, on the contrary, they are alike in unimportant ways and different in important ways, then there is no valid analogy between them and a fallacy of false analogy results. Merely to seize upon some slight similarity as a basis for concluding that what is true of one is also true of the other will usually lead one astray.

Consider this argument, which has been advocated by diverse groups throughout history:

> a) It is necessary to force other people to accept our religious beliefs about an afterlife for their own good, just as force must be used to prevent a delirious person from leaping over the edge of a steep cliff.

Even if we were to grant that our religious beliefs are superior to others, this argument would remain unsound because of the false analogy employed. For in the one case it is a matter of saving a "delirious" person from committing suicide, while in the other case the persons involved are presumably not delirious. Thus it would not follow that, just as force is necessary in the case of the delirious person concerning this life, so force is necessary in the case of other people concerning an afterlife. If someone should say in reply that surely anyone who did not believe in our religion must be mentally incompetent and thus "delirious," that person would commit the fallacy of begging the question. (The mere fact that an individual says other people are delirious is no proof that they are.)

To expose a false analogy—or an *imperfect analogy*, as it is sometimes called—it is necessary to establish that the two things being compared resemble each other in insignificant ways, while they differ in significant ways.

> b)   Why should we sentimentalize over a few hundred thousand native Americans who were ruined when our great civilization was being built? It may be that they suffered injustices, but, after all, you can't make an omelet without breaking a few eggs.

This argument tends to make us wary because of the tasteless nature of its analogy; comparing peoples to eggs and omelets seems a gross oversimplification. The fallacy lies, however, in the comparison drawn rather than in its lack of taste. Even if it were true that it is just as impossible to build great societies without causing pain and suffering as it is to make an omelet without breaking a few eggs, the two cases are not comparable. For to break eggs is not to cause them pain, while to build great empires by destroying people's lives does cause pain.

A similar defect will be found in the following argument from the famous essay "On Suicide" by the eighteenth-century Scottish philosopher David Hume:

> c)   It would be no crime in me to divert the Nile or Danube from its course, were I able to effect such purposes. Where then is the crime of turning a few ounces of blood from their natural channel?

Here again, the things compared differ in a very significant respect: while diverting the Nile or Danube does not destroy the rivers, to divert one of the "channels" (or arteries) of the human body is to do away with it entirely.

In important respects, the analogies in the following arguments are also false:

> d)   What is taught on this campus should depend entirely on what students are interested in. After all, consuming knowledge is like consuming anything else in our society. The teacher is the seller, the student is the buyer. Buyers determine what they want to buy, so students should determine what they want to learn.

The buyer often knows the goods he or she purchases before purchasing them, but does the student know the subject before learning it?

   e) Why should mine workers complain about working ten hours
       daily? Professional people often work at least that long with-
       out any apparent harm.

Where the work is performed, and under what conditions, make
this analogy false.

   Francis Bacon's sixteenth-century argument in support of war
contains an often-cited example of false analogy:

   f) No body can be healthful without exercise, neither natural
       body nor politic; and, certainly, to a kingdom, or estate, a just
       and honorable war is the true exercise. A civil war, indeed,
       is like the heat of a fever; but a foreign war is like the heat
       of exercise, and serveth to keep the body in health; for in a
       slothful peace, both courages will effeminate and manners
       corrupt. *(Of the True Greatness of Kingdoms)*

What is striking about this argument is not that so eminent a
philosopher as Bacon should defend war but that he should either
have failed to note—or expected his readers not to note—that nations are not analogous to persons and that human exercise, unlike war, is not necessarily at the expense of others.

   Metaphors tend to confuse, rather than illuminate, in logic. This
is especially true of analogies, which in essence are expanded
metaphors. We can admire Bacon's metaphorical skill with the
notions of body, exercise, and heat, but we must not mistake his
metaphorical meaning for logical meaning. From the perspective
of logic, Bacon's argument, like that of Hume on suicide, fails to
persuade us because of the false analogy employed.

   On occasion, nevertheless, reasoning by analogy can be instructive. Such a case was the famous discovery by the Greek mathematician Archimedes that a body immersed in fluid loses in
weight an amount equal to the weight of the fluid displaced. Archimedes is said to have made this discovery while attempting to
solve a problem for King Hieron, who wished to know what metals
had been used in his crown but did not wish to destroy the crown
by melting it down. Archimedes solved the problem by observing

that the water in his bath rose as his body displaced it. He reasoned by analogy that a certain weight of gold would displace less
water than the same weight of silver because it was smaller in
volume. He tested the crown and found that it was made of impure
gold.

A story about the great Renaissance astronomer Copernicus be-
longs to the same category. Drifting in his boat near the bank of
a river one day, Copernicus had the illusion that the bank was
moving while his boat remained stationary. It suddenly struck the
astronomer, it is said, that the same illusion might be the cause of
the belief that the earth remains still while the sun moves around
it. Reasoning by this analogy, Copernicus revolutionized human
understanding of the universe and founded modern astronomy.

More often, however, discoveries are made by testing analogies
carefully, for many of those used in argument will be found to be
fallacious. We smile today at our forebears who delegated the
sowing of crops to women who had borne many children, on the
assumption that human fertility was somehow analogous to a rich
harvest. We wonder how a superstition could arise which led people, when wishing to injure an enemy, to make an image of the
enemy and then destroy it. But we moderns are misled by analogies every day. The best defense against this distortion of facts is
always to sort out those aspects of the analogy that are relevant to
an argument and those that are not.

9. THE FALLACY OF FALSE CAUSE

The *fallacy of false cause* is an argument which suggests that
events are causally connected when in fact no such causal connection has been established. Although formerly widespread, this fallacy does not appear today in the crude forms it once assumed, due
largely to the impact of general education. Most of us no longer
take the trouble to walk around a ladder rather than under it—
though a hotel will seldom risk naming the thirteenth floor by that
number, for fear that no one would want to sleep there.

By the same token, we have no doubt become too sophisticated
to take seriously the scheme of the nineteenth-century English
reformer, Thomas Malthus, who, noting that sober and industrious farmers owned at least one cow while those who had none

were usually lazy and drunken, recommended that the government give a cow to any farmers who had none, in order to make them sober and industrious.

Yet the following arguments containing the fallacy of false cause often persuade people today.

a) Since every major war in which we have taken part during the last few generations has happened when we had a Democratic president, we ought therefore to think twice before voting for a Democrat in this presidential election.

b) More and more young people are attending high schools and colleges today than ever before. Yet there is more juvenile

"Wait a minute, Vince! Last summer —
remember? Some little kid caught you,
handled you, and tossed you back in the
swamp ... *That's* where you got 'em."

The fallacy of the false cause. Any number of other factors or events in the recent or distant past might have contributed to this creature's unfortunate condition. (This cartoon is also, obviously, a play on an older and even more absurd fallacy—the idea that people can get warts from handling a toad.)

delinquency and more alienation among the young. This makes it clear that these young people are being corrupted by their education.

The fact that two events occur at roughly the same time as one another is distorted into an assumption that one event is the cause of the other.

Glendower, a character in Shakespeare's *King Henry IV, Part I,* makes a similar mistake, in assuming that there was something special about his birth.

> c) GLENDOWER:
>
> > At my nativity
> > The front of heaven was full of fiery shapes,
> > Of burning cressets: and at my birth
> > The frame and huge foundation of the earth
> > Shaked like a coward.
>
> HOTSPUR:
> Why so it would have done at the same season
> if your mother's cat had but kittened, though
> yourself had never been born.
> (act 3, sc. 1)

Here, Hotspur's reasoning is better than Glendower's.

Another common form of the fallacy of false cause is the mistaken assumption that because one event occurred prior to another event, it therefore caused the second event.* Actually, every given event is preceded by a multitude of events, any one of which can be its cause. Sequence alone is no proof of consequence.

---

*The widely used Latin phrase for this fallacy is *post hoc, ergo propter hoc,* meaning, literally, "after this, therefore because of this."

Failure to heed this simple truth gives rise to the fallacies in this argument:

> d) Twenty-five years after graduation, alumni of Harvard have an average income five times that of people of the same age who have no college education. If a person wants to be wealthy, he or she should enroll at Harvard.

Here the investigation of causes has failed to proceed far enough. Although attending a school such as Harvard no doubt contributes to the kind of income one is likely to make, it is well to remember that Harvard attracts and accepts only outstanding students, or students who come from a background of affluence. Harvard alumni, therefore, would probably achieve high incomes regardless of which college they attended—perhaps whether they attended college at all.

Analysis of some instances of false cause reveals that two events may be related even though neither is the cause of the other. In such instances, both are effects of a third event, which is the cause of each of them. An interesting historical example concerns the ibis, a bird sacred to ancient Egypt. Egyptians worshiped the ibis because each year, shortly after flocks of ibis had migrated to the banks of the Nile river, the river overflowed its banks and irrigated the land. The birds were credited with causing the precious flood waters when in fact both their migration and the river's overflow were effects of a common cause, the change of season.

Neither immediate temporal succession or more remote temporal succession is sufficient for establishing causal connection. The fact that *homo sapiens* follows the ape in the succession of primates is no proof that we are descended from the ape; nor is the fact that the Roman Empire declined after the appearance of Christianity proof that Christianity was the cause of its decline. This passage from Mark Twain's *Huckleberry Finn* illustrates how far astray such causal analyses can go:

> e)   I've always reckoned that looking at the new moon over your left shoulder is one of the carelessest and foolishest things a body can do. Old Hank Bunker done it once, and bragged about it: and in less than two years he got drunk and fell off of the shot-tower and spread himself out so that he was just kind of a layer, as you may say; and they slid him edgeways between two barn doors for a coffin, and buried him so, so they say, but I didn't see it. Pap told me. But anyway it all come of looking at the moon that way, like a fool. (ch. 10)

Huck's reasoning here is a classic example of *post hoc, ergo propter hoc*, the root of much prescientific superstition.

The form in which people commit the fallacy of false cause has tended to change with the progress of science. The notion that nature acts with a purpose is slowly becoming a thing of the past. It would be a rare person who would argue today that it will be a hard winter because hollyberries (which nature provides for birds in hard weather) are abundant this year. We find it quaint that people once explained why a container filled with water breaks when the water freezes by assuming that water contracts in freezing, creating a vacuum which nature "abhors." We should have thought that simple observation would have shown them that water expands rather than contracts in freezing and that it is the expansion that breaks the container.

The example of the freezing water comes from a logic textbook published in 1662 by the French Renaissance philosopher Antoine Arnault. Enormously popular in his day, Arnault's *The Art of Thinking* went through numerous editions at a time when modern science was being born. Arnault took pains to identify errors in earlier scientific thinking, as in the following passage:

> The sophism* of taking for a cause what is not a cause may occur in still another way. If we argue that since one event

---

*A *sophism* is similar to a fallacy. Like our word *sophistication,* the term *sophism* derives from the Greek *sophos,* meaning "clever." A sophism is an argument that though correct in appearance is nevertheless invalid. Prior to Aristotle's founding the science of logic, some of the philosophers of Greece were called *sophists.* Logic had become a mere game to many of them, and their reasoning was often false. *Sophist* thus came to mean one who deceives, and *sophism* a deceptive argument. By contrast to a sophism, a fallacy may result from simple error rather than from a wish to deceive.

---

occurs after another then the latter event must be the cause of the former, we commit the sophism in the form called *post hoc, ergo propter hoc.* Reasoning in this way people have concluded that the Dog Star is the cause of that extraordinary heat we feel during the dog days. Virgil, writing about the Dog Star, called Sirius in Latin, said:

> *Even as fiery Sirius:*
> *Bearer of drought and plague to feeble man*
> *Rises and saddens the sky with baleful light.*
> *Aeneid* X:273–275

But Gassendi has very correctly observed that nothing could be less likely than crediting the Dog Star with the heat of August. The Dog Star's influence ought to be strongest in the region to which the star is closest. But in August the Dog Star is much closer to the region below the equator than to us: and yet while we are in the dog days, the regions below the equator have their winter season. So, inhabitants below the equator are more justified in thinking that the Dog Star brings cold than we are for thinking that the star brings the heat. (Trans. James Dickoff and Patricia James. Indianapolis: Bobbs-Merrill, 1964, p. 255)

Arnault here traces the origin of our modern idiom *dog days* for hot weather, in the course of his explanation of how early scientists committed the error of *post hoc. . . .*

A more recent example of erroneous causal analysis occurred in the modern treatment of schizophrenia. Dr. Manfred Sakel discovered in 1927 that schizophrenia can be treated by administering overdoses of insulin, which produce convulsive shocks. Hundreds of psychiatrists drew a faulty conclusion and began to treat schizophrenia and other mental disorders by giving patients electric shocks without insulin. At a psychiatric meeting some years later, Dr. Sakel sadly came forward to explain that electric shocks are actually harmful, while insulin treatment restores the patient's hormonal balance. The doctors had confused a side effect with a cause. In general, however, we are growing more sophisticated in our reasoning about natural or physical causes.

This is not so true, on the other hand, of our reasoning about psychological causes. Many people still seem to believe, for example, that merely saying an event will happen helps make it happen. A good case can be made for this form of the "self-fulfilling prophecy," but on psychological rather than magical grounds. What actually happens is that, for example, fearing others will be unfriendly to us, we act in an unfriendly manner that frequently makes them unfriendly in return. Or, expecting a friendly welcome, we greet others warmly and they respond in kind.

The belief in bad luck or a "jinx" is similar. Here too, what actually happens is that people who believe they have been "jinxed" are likely to falter in ways that will work against them. They may, on the other hand, take special pains to function well, in which case they are likely to experience good fortune. There are causes at work here, but they are often causes other than those we think. As in all cases of the fallacy of false cause, the best line of reasoning is one that distorts the facts as little as possible.

Some logicians have identified still another form of this fallacy, which is called *circular cause*. The fallacy, which resembles one of the forms of the fallacy of begging the question, results when the cause of a phenomenon is attributed to one of two cooperative causes that produce an event. To say, for example, that one runs because one is afraid and that one is afraid because one runs, without further explanation, involves circularity. To argue further, for example, that the country is poor because of a depression, and the country is in a depression because the people are poor, is another such case of circular causation. The following, from Antoine de Saint Exupéry's classic *The Little Prince,* is a further example of the fallacy:

> The next planet was inhabited by a tippler. This was a very short visit, but it plunged the little prince into deep dejection.
>
> "What are you doing there?" he said to the tippler, whom he found settled down in silence before a collection of empty bottles and also a collection of full bottles.
>
> "I am drinking," replied the tippler, with a lugubrious air.
>
> "Why are you drinking?" demanded the little prince.
>
> "So that I may forget," replied the tippler.
>
> "Forget what?" inquired the little prince, who already was sorry for him.
>
> "Forget that I am ashamed," the tippler confessed, hanging his head.
>
> "Ashamed of what?" insisted the little prince, who wanted to help him.
>
> "Ashamed of drinking!" The tippler brought his speech to an end, and shut himself up in an impregnable silence.
>
> And the little prince went away, puzzled.
>
> "The grown-ups are certainly very, very odd," he said to himself, as he continued on his journey.

10. THE FALLACY OF SLIPPERY SLOPE

An interesting variant of both the fallacies of false analogy and false cause is slippery slope. Two other names that fit this fallacy are *bad precedent* and *thin end of the wedge.* This argument objects to a proposal or a position on the ground that, although not in itself bad or dangerous, it will lead to a situation that is. The following, prompted by a court ruling which allowed the University of Georgia to demand to know how professors on a tenure committee voted, is a typical example of it:

a) The ruling sounds like the beginning of a totalitarian state. Judge Owens is violating the concept of the secret ballot by demanding that faculty members reveal how they voted. Next the government will want to prohibit secret voting in unions, professional organizations, civic organizations, corporations, and finally in the general elections.

The writer urges that the ruling be rejected because it would set a bad precedent and threaten to set off an undesirable chain of events. But the writer's reasoning is faulty, for he or she imagines that all these cases of voting are similar and that therefore disclosure in one will ultimately force disclosure in the others as well. But are these cases and the issues they are concerned with as similar as the writer supposes? If they are not, they need not lead to the same results. The fear of stepping on a slippery slope here is therefore groundless.

This type of argument was popular during the Vietnam War. Indeed, hardly a day passed without a letter in some newspaper expressing the fear that if Vietnam fell, Cambodia would follow, then Thailand, Burma, India, and so forth. And similar fears were raised during the Falklands conflict, as in the following letter:

b) It is asinine to talk of mediating between Britain and Argentina. What is needed is strong U.S. military action to help Britain retake the Falklands and to punish Argentina. Else we shall soon be losing Guantanamo, Guam, Virgin Islands and Catalina.

Here, too, the writer seems to imagine we are on a slippery slope and that if we take one step on it we will not be able to stop and will slide down the whole slope. But stop we often can, for most things are not like slippery slopes and do not lead to the envisioned dire consequences. Each new situation as it arises can be evaluated anew and decided on its merits. If the things were indeed similar, they would lead to similar effects. The cases, therefore, need always be examined individually to see how similar they are.

The fear that we may not be able to stop once we embark on such a slope is not a new one. When in 1698, for example, psalm books began to be printed with proper musical notation, a writer in the *New England Chronicle* remarked as follows about this new method of "singing by rule":

c) Truly I have a great jealousy that if we once begin to sing by rule, the next thing will be to pray by rule, and preach by rule; and then comes popery.

Nor is this fallacy and the fear it tries to conjure up likely to be abandoned soon. The following appears in a recent book on nutrition:

d) If the "experts" decide today that we should have fluorides in our tea, coffee, frozen orange juice, lemonade, *and every cell of our bodies,* what's in store for us tomorrow? What about vitamin C in the water, considered by some to be much more important than fluoride? What about tranquilizers to avoid civil disorders? What about birth-control chemicals to be routed to the water in certain ethnic neighborhoods? When the time comes, of course, you can be sure it will be done for "your comfort and safety." (David Reuben, *Everything You Always Wanted to Know about Nutrition*)

Again, we need to ask ourselves whether the situations mentioned here are comparable. Does our accepting the use of fluorides and vitamin C commit us inevitably to accepting tranquilizers and birth-control chemicals? Are there no differences between these situations? Should they arise, will we not be able to argue that while the use of fluorides and vitamin C are acceptable since they are likely to enhance the life and health of all concerned, this is hardly the case of the use of the other substances?

Such arguments are only convincing by ignoring pertinent differences between cases falsely assimilated.

11. THE FALLACY OF IRRELEVANT THESIS

A *thesis* is a position that one advances by means of an argument—as in a master's thesis, in which a particular view of a subject is set forth with supporting evidence. In logic, a thesis can be equated with a conclusion. The *fallacy of irrelevant thesis,* therefore, is an argument in which an attempt is made to prove a conclusion that is not the one at issue. This fallacy assumes the form of an argument that, while seeming to refute another's argument, actually advances a conclusion different from the one at issue in the other's argument.

Of all the fallacies studied thus far none is potentially more

"Hey, I'm not *crazy* .... sure, I let him
drive once in a while, but he's never,
never off this leash for even a second."

The man is presenting an irrelevant thesis. His compliance with city leash laws is not being questioned; it has little to do with the dog's abilities as a driver, or the man's mental competence.

deceptive—or, for that matter, more interesting—than irrelevant thesis. This fallacy goes by a variety of names: *irrelevant conclusion, ignoring the issue, befogging the issue, diversion,* and *red herring. Red herring* may seem a puzzling name. It derives from the fact that escapees sometimes smear themselves with a herring (which turns brown or red when it spoils) in order to throw dogs off their track. To sway a red herring in an argument is to try to throw the audience off the right track onto something not relevant to the issue at hand.

The fallacy of irrelevant thesis derives its persuasive power from the fact that it often does prove a conclusion or thesis (though not the one at issue). An example will make this clear.

> a)   The advocates of conservation contend that if we adopt their principles we will be better off than if we did not adopt them. They are mistaken, for it is easy to show that conservation will not produce an Eden on earth.

Two quite different questions are clearly at issue here: (1) whether conservation is the best available measure, and (2) whether conservation will produce a Garden of Eden on earth. By refuting the second argument rather than the first, argument *a* commits the fallacy of irrelevant thesis. The argument seems persuasive, at first, because it points out what we feel must be true: no conservation measures can assure us of an Eden. We may therefore be tempted to reject out of hand any suggestion that conservation is worthwhile. But this would be a mistake, for all that is at issue is whether conservation is our best alternative.

The famous philosophical novel *Candide* by the eighteenth-century French thinker Voltaire yields another example of irrelevant thesis. Voltaire invented the character Dr. Pangloss, who is tutor to the hero Candide, to represent Leibniz, a philosopher with whose optimism Voltaire disagreed.

> b)   Dr. Pangloss contends that this world is the best of all possible worlds which God could have made. What a ridiculous assertion! As if everything in this world were as good as it could be!

Here too, by misrepresenting the point at issue, the refutation seems to appear cogent. That it is not at all cogent becomes obvious when we realize that Leibniz did not say that everything in this world is as good as it can be, but only that this world is better than some other worlds God could have made. (God, Leibniz would have argued, could have made a world that would have been free of, say, earthquakes—but then something else would have had to be wrong with it, something worse than earthquakes. All things considered, Leibniz maintained, this is the best "possible" world God could have made.)

These two examples illustrate well a maneuver frequent in argument. It consists of imputing to one's adversaries opinions a good deal more extreme than those they have set out and are willing to defend. Distorting the position in this way makes it appear ridiculous and thus easily overthrown. If the adversaries are tricked into defending a position that is more extreme than their original one, they are in all likelihood destined to fail. Although this is a popular trick in debating, it is a dishonest one.

Not all fallacies of irrelevant thesis stem from a conscious effort to distort, however. Pressed for time, or lacking distance, we may regard certain facts as relevant that we would otherwise recognize as beside the point. No bad faith was involved, for example, in this statement with which the great French philosopher René Descartes opened his profound *Discourse on Method:*

> c) Good sense is of all things the most equally distributed among men. For everybody thinks himself so abundantly provided with it that even the most difficult to please in all other aspects do not commonly desire more of it than they already possess.

The statement is designed to encourage his readers to exert their best efforts—just as Descartes supposedly did when, with no more common sense than other people, he discovered a method which he now shares with the world. From a logical point of view, however, this argument is fallacious. Although it may be true that everyone is pleased with the amount of good sense he or she possesses, this is no proof at all—and is irrelevant to the question—that everyone possesses as much good sense as everyone else.

The same can be said of the following argument, which, again, was probably not intended to distort the facts but which nevertheless does so:

d) I fail to see why hunting should be considered cruel when it gives tremendous pleasure to many people and employment to even more.

With a little distance, we cannot fail to see that whether hunting gives employment or pleasure to people is irrelevant to whether it is cruel to animals. If that is irrelevant, then the argument that hunting is cruel has yet to be challenged. Not to do so is to distort the issue. As in the other cases we have examined, this argument avoids what it is supposed to disprove (that hunting is cruel to animals) and proves what has not been argued (that hunting is advantageous for people).

Just as some people faced with a problem simply get up and run away, so in argument some people faced with a difficult or unpleasant line of reasoning simply take cover under some piece of irrelevance. Typical is the prosecutor who tries to persuade a jury of the defendant's guilt by arguing that murder is a horrible crime.

Because nothing is so effective in relaxing people's attention as laughter, it is not surprising that fallacies of irrelevant thesis can be amusing, as in this tale about the theft of a pig:

e) "Well now, Patrick," said the judge. "When you are brought face to face with Widow Maloney and her pig on Judgment Day, what account will you be able to give of yourself when she accuses you of stealing the little animal?" "You say the pig will be there, Sir?" said Pat. "Then I'll say: 'Mrs. Maloney, there's your pig!' "

This argument is humorous, but the same fallacy can be deadly serious, as Susan Stebbing points out in her analysis of a British parliamentary debate over the exportation of guns, as compared with the exportation of children's firecrackers:

f) SIR PHILIP: You do not think that your wares are any more dangerous or obnoxious than boxes of chocolates or sugar candy?
SIR CHARLES: No, or novels.
SIR PHILIP: You don't think it is more dangerous to export these fancy goods to foreign countries than, say, children's crackers?

> SIR CHARLES: Well, I nearly lost an eye with a Christmas cracker, but never with a gun.

As Stebbing points out, Sir Charles's response diverts attention away from the point at issue by use of a joke. "The hearer might willingly assent to the suggestion that someone might 'nearly lose an eye with a Christmas cracker' although he has never been in danger from a gun. Crackers, however, are not made for this purpose, whereas armaments are made solely for the purpose of killing and wounding people and destroying buildings. But it is *armaments* that are being discussed" (*Thinking to Some Purpose.* Harmondsworth: Penguin Books, 1939, p. 196).

All such befogging of the issue is best rebutted by the simple statement: "True, perhaps, but irrelevant." With presence of mind, a simple, incisive retort such as this will bring an audience back to the issue and may even succeed in demolishing the offender. The story is often told of an actor, playing the role of Shakespeare's King Richard III, who had just done his best with the line: "A horse! A horse! My kingdom for a horse!" "Will an ass do?" came a shout from the gallery. "Certainly," the actor shouted back. "Come on down." The shift of attention that lies at the heart of the fallacy of irrelevant thesis can be used effectively to demolish such a fallacy.

When diverting attention, an appeal to emotion can be as effective as humor.

> g)   The president's decision to veto the tax bill was a wise one. Never has a man taken office under such difficult conditions. The nation's economy, dislocated by a long and costly war, its nerves stretched to the snapping point by the threat of another, is in desperate need of stewardship. The president has made this decision faced with revolt within his own party and with a torrent of abuse from a hostile press.

This argument attempts to divert attention from the issue by an appeal to sympathy. As such, it verges on a type of fallacy which forms a group of its own and which will be discussed in the following chapter.

EXERCISES

Identify the fallacy—false analogy, false cause, slippery slope, or irrelevant thesis—that is committed in or that could result from each of the following. Explain the error committed in each case.

* 38. I am the father of two daughters. When I hear this argument that we can't protect freedom in Europe, in Asia, or in our own hemisphere and still meet our domestic problems, I think it is a phony argument. It is just like saying that I can't take care of Luci because I have Lynda Bird. We have to take care of both of them and we have to meet them head on. (Lyndon Johnson)

* 39. I do not permit questions in my class, because if I allow one student to ask a question, then everyone starts asking questions and the first thing you know, there is no time for my lecture.

40. Student government is a mistake from the beginning. Look what happens in homes where parents let the children run things their own way!

* 41. Television can't be harmful to children, because it occupies their attention for hours and thus keeps them off the street.

* 42. An electrical power failure in the Southwest is imminent. UFOs have been seen over Boulder Dam, just as they were seen over the Niagara powerhouse shortly before the great New England blackout.

43. Early in 1950 the United Nations International Children's Emergency Fund distributed powdered milk to expectant mothers in the village of Polykastron in Greece. Shortly afterward, two women who used it gave birth to twins on the same day—the first twins born in the village in ten years. The women of Polykastron drew the obvious conclusion: they decided they'd rather not use UN powdered milk.

* 44. "No man is an island"—Much as we may feel and act as individuals, our race is a single organism, always growing and branching—which must be pruned regularly to be healthy. This necessity need not be argued; anyone with eyes can see that any organism which grows without limit always dies in its own poisons. The only rational question

is whether pruning is best done before or after birth. (Robert A. Heinlein, *The Notebooks of Lazarus Long*)

45. How can nuclear energy be so bad if it is our best source of energy?

46. Grove City College has not taken a penny in direct government aid since it was founded 103 years ago. No matter. The Presbyterian-affiliated school has been showered with federal decrees and directives, most of which it has ignored. One form from the Department of Health, Education and Welfare was not so easily tossed aside, however. HEW has been pressing Grove City since 1977 to sign a confirmation that it is complying with Title IX of the Education Amendments of 1972, which bans discrimination against women. Grove City officials say the college has no quarrel with women's rights. But because the college gets no direct government funds, it has refused as a matter of principle to sign the form. To college President Charles MacKenzie, the issue is academic freedom. "Once the nose of the camel gets in the tent, the whole camel moves in," he said in an interview at the school's campus in western Pennsylvania's rolling hills. "If we signed this, we'd be expected to sign compliance forms for everything under the sun." (News item)

*47. Why do I love my country? Two hundred words would not be enough to cover all the reasons why I feel the way I do about my country. In my eighty-one years of living I have found so much evidence to bolster my belief in the promise set forth for me by my mother, who had come to America from Sweden when she was very young, to become a true American, patriotic and unquestioning. Through a long life of happiness, intermingled with some tragedy and sorrow, there was never any wavering in that patriotic zeal. She loved her country and never blamed it for any of the bad. Rather, she gave her country credit and thanks for all the good. I believe that was right. Oh, I am not blind to things like Watergate. . . . There is much that is wrong in man's world. But the ideals and principles on which our country was founded still make up the cornerstone of its structure and there are many men and women who will sincerely work to keep them there. My part in all of it is to keep on working, too. Yes, I love my country. (Letter to the editor)

48. I'm tired of people asking me to remove my dark glasses so that they can see my blue eyes. I wonder if I would have been as successful if I had been born with brown eyes. (Paul Newman)

49. If architects want to strengthen a decrepit arch, they *increase* the load that is laid upon it, for thereby the parts are joined more firmly together. So, if therapists wish to foster their patients' mental health, they should not be afraid to increase that load through a reorientation toward the meaning of one's life. (Viktor E. Frankl, *Man's Search for Meaning*)

50. QUESTION: Is stealing sometimes justifiable?
    ANSWER: In baseball it is even honorable.

51. I look upon Fran's place (legal brothel in Nye County, Nevada) as a kind of insurance policy for the safety of young women and children in this area. As long as there is a kind of safety-valve sort of a place where men can go to, my fifteen-year-old daughter Hillary can walk home at night in perfect safety.

52. At the Los Angeles meeting of the United Auto Workers, Senator Edward M. Kennedy once more called for national health insurance. If socialized medicine will result in better and lower-cost health care, shouldn't the same logic be applied to automobiles? Wouldn't nationalization of the auto industry produce better and lower-priced cars and nationalization of auto mechanics and garages produce higher-quality, less expensive repairs? (Letter to the editor)

SUMMARY

This chapter has presented three categories of presumptive fallacies, arguments which smuggle in unfounded or unproven assumptions under the guise of valid argument forms.

In the first type of presumptive fallacy, as we have seen, the error lies in overlooking essential facts. Three fallacies of this type were examined.

Sweeping generalization was shown to result when a generalization is applied to a special case which properly falls outside of it, as when horseback riding is recommended for someone with a heart condition.

Hasty generalization, on the other hand, was shown to be the opposite of sweeping generalization. Here, an isolated or exceptional case is used erroneously to support a universal conclusion, as when a bad experience with a former husband is used to prove that all men are no good.

Bifurcation, the third fallacy of this type, overlooks the far-flung possibilities that lie between two polar alternatives, as in the assertion that, if you know about BMW, you either own one or you want one.

In the second type of presumptive fallacy, the error was shown to stem from an evasion of the facts. Here we examined four fallacies.

Begging the question was shown to result from assuming in the premises of an argument the very conclusion that the argument is supposed to prove. An example cited was the assertion that bodies fall because they have a downward tendency.

Question-begging epithets were shown to be similar to fallacies of begging the question in that they too affirm something that is not yet proved. Question-begging epithets do so by means of slanted language, as in the use of the epithet "an effete corps of impudent snobs" to characterize dissenters.

Complex question was shown to assume a certain answer to a prior question that was never asked. An example cited was the complex question: Did John ever give up his bad habits?

Special pleading, the last of the four fallacies that evade facts, was seen as an attempt to set up a double standard—a special standard for ourselves and another standard for others. This fallacy evades the facts by being prejudiced in favor of one's own side, as when we refer to our troops as *devoted* while those of our enemy are called *fanatical.*

The third and last type of presumptive fallacy dealt with four fallacies that distort the facts.

False analogy was shown to distort by making the facts under discussion appear more similar to another set of facts than they really are. An example cited was Bacon's comparison of a nation waging war with a person engaging in exercise.

False cause was shown to distort facts by assuming that two events are causally connected when in fact they may not be. The Egyptians' worship of birds that migrated to the Nile valley just before a flood was cited as an illustration.

Slippery slope was shown to distort by making it appear that a certain position would set off an undesirable set of events when it is far from certain that this would be the case. An example cited was David Reuben's argument against the use of flourides.

Irrelevant thesis, the last fallacy of presumption discussed here, was seen as a distortion of facts by means of the substitution of another issue for the one actually in dispute. We examined in this connection the argument that, since conservation cannot guarantee us an Eden on earth, we should not practice conservation as the best available course.

EXERCISES

Identify the fallacy of presumption that is committed in or that could result from each of the following. Explain the error committed in each case.

53. It's ridiculous for Professor Ames to devote all of her time to Egyptology. If everybody worried about ancient Egypt the world would soon grind to a stop.

*54. Since education results in improvement, everyone, irrespective of ability, should be given the advantage of a higher education.

55. I have never known anybody named Jim who wasn't nice.

*56. You can tell that Frank is a disreputable person by the character of his associates, because people who go around with somebody like Frank are the lowest type.

*57. Why isn't a nice person like you married?

*58. Marijuana can't be all that bad. Everyone knows about barroom brawls, but marijuana makes people peaceful.

*59. I think his daughter's marriage must have worried him dreadfully. She was his only child, you know. He never talked about her but I noticed that his hair began to turn white after the wedding.

*60. The United States had justice on its side in waging this war. To question this would give comfort to our enemies and would therefore be unpatriotic.

*61. The deplorable deterioration of governmental efficiency one finds here is a direct cause of a widespread indifference on the part of the people with respect to governmental affairs today.

* 62. High retail prices are beneficial, for they stimulate employment, increase agricultural production, and stimulate the development of new products.

* 63. I don't care if he did weigh three times as much as you. A good scout always tries to help. You should have jumped in and tried to save him.

64. American workers shouldn't kick. They're much better off than their European counterparts.

* 65. America: Love It or Leave It.

66. Of course she loves me. She told me that she loved me, and I believe her! Would she lie to someone that she loved?

* 67. That the world is good follows from the known goodness of God; and that God is good is known from the excellence of the world he has made.

* 68. We should legalize gambling in this state because, first, it would provide a rich new source of revenue; second, it would encourage tourists to come and spend their money here; and third, it would cost us nothing to get these new moneys, just the passing of a law.

69. A parricide is in the same relation to his father as a young oak to the parent tree, which, springing up from an acorn dropped by the parent, grows up and overturns it. We may search as we like, but we shall find no vice in this event. Therefore there can be none in the others where the relations involved are just the same. (David Hume, *Treatise of Human Nature*)

* 70. Rosenbaum sold hot dogs from a pushcart.
"How's business?" asked an acquaintance.
"Could be worse!" said Rosenbaum. "I put away already two thousand dollars in the bank!"
"That's good," said the friend. "Maybe you could lend me $5."
"I'm not allowed."
"What, not allowed?"
"I made an agreement with the bank. They agreed not to sell hot dogs if I promised I wouldn't make loans."

* 71. "I have been tracking Mr. Feinstein's advice for one year. When he wrote that a certain set of stamps were going to increase in value, they did. This and other suggestions,

they all went up. So if he claims that certain banknotes are going to rapidly increase in value, bank on it." (Testimonial from an investment pamphlet entitled "The Treasure of Alan Shawn Feinstein")

* 72. One of the great rabbis of the last hundred years was riding on a train in Russia, and he overheard a conversation between a Christian missionary and some deeply religious but uneducated Jews. The Jews had just expressed their confidence in the judgment of the ancient rabbis concerning the Messiah. "In that case," asked the Christian, "how can you explain the fact that Rabbi Akiva (one of the greatest Talmudic rabbis) initially thought that Bar Kochba (a Jewish revolutionary of the second century) was the Messiah?" The Jews were taken aback and could find no answer. The rabbi, who had been listening quietly, turned to the Christian missionary and asked, "How do you know that Bar Kochba wasn't the Messiah?" "That's obvious," he replied; "Bar Kochba was killed without bringing the redemption."

* 73. California obstetrician William Waddill stood trial in 1978—and again in 1979—for allegedly strangling a baby girl delivered alive after he performed a saline abortion on Mary Weaver, age eighteen. Waddill admitted that the thirty-one-week-old fetus was struggling for breath, but claimed that they were dying gasps and that "no doctor walking on the face of this earth could have resuscitated that baby." (News item)

  74. For almost a decade, the gruesome livestock killings have baffled law-enforcement officials. Hundreds of cattle in twenty-seven states have been found dead and mutilated— their ears, lips, tongues, eyes and genitals cut out with almost surgical precision. The mutilations have given rise to wild rumors about flying saucers and witchcraft. . . . The State Department of Criminal Investigations (DCI) says it finally has some suspects—members of certain unnamed "satanic groups." . . . The DCI won't disclose the names of the groups under suspicion, but one has been ruled out— the Iowa Witches Association. According to DCI director Gerald Shanahan, two witches from the association came forward to say that their organization consists only of good witches and had no part in the mutilations. "They must be good witches," says Shanahan, "because they provided investigators with leads and information." (News report)

75. Why are you losing your hair?
    I worry a lot.
    What do you worry about?
    Losing my hair!

76. As Ernest Hemingway noted, the only thing that sets rich people apart from all others is that they have more money. This was also the opinion of Joseph P. Kennedy, father of the noted political clan. When a Princeton University psychology student sent him a long, involved questionnaire in an attempt to probe the personality characteristics of the rich, Mr. Kennedy returned the questionnaire unanswered with a note stating, "I am rich because I have a lot of money."

77. There are psychiatric patients who are said to be incapacitated by having more than one self. People like this are called hysterics by the professionals, or maybe schizophrenics, and there is, I am told, nothing much that can be done. Having more than one self is supposed to be deeply pathologic in itself, and there is no known way to evict trespassers. I am not sure that the number of different selves is in itself all that pathologic; I hope not. Eight strikes me personally as a reasonably small and easily manageable number. . . . Actually, it would embarrass me to be told that more than a single self is a kind of disease. I've had, in my time, more than I could possibly count or keep track of. The great difference, which keeps me feeling normal, is that mine (ours) have turned up one after the other, on an orderly schedule. (Lewis Thomas, "Living with One's Selves")

*78. Your mother is going into the hospital for elective surgery and really wants you to be at her side. You were planning to spend the weekend struggling with a new campaign due at work on Monday. Do you say yes or no?

*79. I believe that every man who has ever been earnest to preserve his higher or poetic faculties in the best condition has been particularly inclined to abstain from animal food, and from much food of any kind. It is a significant fact, stated by entomologists, I find it in Kirby and Spence, that "some insects in their perfect state, though furnished with organs of feeding, make no use of them," and they lay it down as "a general rule, that almost all insects in this state eat much less than in that of larvae. The voracious caterpillar when transformed into a butterfly . . . and the gluttonous maggot when become a fly," content themselves with a drop or two of honey or some other sweet liquid. The abdomen under the wings of the butterfly still represents the larva. This is the tidbit which tempts his insectivorous fate. The gross feeder is a man in the larva state; and there are whole nations in that condition, nations without fancy or imagination, whose vast abdomens betray them. (Henry David Thoreau, *Walden*)

80. Forced military service is an issue that touches our fundamental concepts of liberty and rights. To argue in support of the draft is to concede to abrogation of inalienable rights. To persist in such support after so recent a slaughter of draftees in Vietnam is to confess a vicious moral turpitude.

81. In the good old days of murder, assault, and battery, you usually got yours for good reason and usually by someone you knew. Nowadays, though, it's not at all uncommon to get terminally creamed by a stranger for occupying a disputed parking space or having brown hair. I think circumstances have overtaken such simple remedies as gun control. I propose, instead, a gun distribution program, a return to the law of the old West, as it were. Give weapons to each able-bodied man, woman, and child, together with fast-draw instructions, and then let them rip. Only then will anonymous creeps reconsider their senseless attacks, because the next time they take aim, they just might find themselves looking down the barrel of the new Gary Cooper. (Letter to the editor)

82. And a phenomenon began that in my opinion has not yet been properly evaluated: the decline of the horse as a symbol of power and strength in wartime. In the fifth century B.C. (a period called "time-axis" by Karl Jaspers, meaning that several events of historical moment occurred in different parts of the world at the same time), a number of great philosophies and religions were born simultaneously. According to Jaspers, this was due to the sudden appearance of horse-riding peoples in the civilized world, which facilitated the diffusion of ideas on a vast scale. If such is the case, I then say that the opposite also applies and that the

gradual decline in importance of the horse in the epoch we are now discussing [the seventeenth century—the age of genius] also has deep significance. Man began to dismount, and it is my belief that the man on foot can always think better and more democratically than the man on horseback. (Félix Marti-Ibáñez, *Tales of Philosophy*)

83. This radical plot has been hatched by un-American agitators.

84. I completely disagree with Smith's assertion that to really conserve gasoline we need to "tax the hell" out of it. Increasing the cost of gasoline and diesel fuel will only set off another round of price-wage increases, which will pound another nail in the coffin of the American free enterprise system. More taxes on petroleum products, whether they are directly on the oil companies—the so-called windfall profits tax—or charged to the consumer at the pump, will not produce one more barrel of oil. (Letter to the editor)

85. Which would you rather be, right or happy?

*86. The anger and frustration in C. Jay Kim's voice came as a surprise. It was supposed to be a sedate affair, a dinner meeting of about twenty leaders of the Korean community at a restaurant on Olympic Boulevard. But Kim was so mad he barely had time to glance at his food. "What are we supposed to do?" he asked. "Asians work hard and don't complain very much, so now they will tell us we're not a minority. We can't get the same kind of help the blacks and Chicanos get.... What are we supposed to do to be a minority? Drink wine all day and stay on welfare?" (News item)

87. If we really believe in the right to life, liberty, and the pursuit of happiness, we'd have abolished prisons long ago.

88. I see that our courts are being asked to rule on the propriety of outlawing video games as a "waste of time and money." It seems that we may be onto something here. A favorable ruling would open the door to new laws eliminating show business, spectator sports, cocktail lounges, the state of Nevada, public education, and of course, the entire federal bureaucracy.

Find examples from your daily reading of the fallacies discussed in this chapter and explain in detail why you think they are fallacious.

ANSWERS TO STARRED EXERCISES

1. *Hasty Generalization:* Much more is involved in being "a great writer" than simply the skill in constructing original plots. More important is what one is able to do with the story constructed—and here no one has surpassed Shakespeare (even if, as is true, he did not make up his own plots).

3. *Bifurcation:* The terms *good* and *cheap* are not necessarily incompatible. It is entirely possible to find a cheap car that is good—that is, that functions well, does not cost much to maintain or operate, and has other such characteristics. The speaker implies that the terms are contradictory (that it is impossible for both to be true and also impossible for both to be false) when in fact they are not.

4. *Hasty Generalization:* When one considers how many doctors there are and how many diagnoses are made in the course of a month, three such faulty diagnoses hardly justify the general conclusion that all doctors are alike or that they do not know any more than we do.

6. *Bifurcation:* It is true that the wealthy sometimes use their wealth to try to influence legislators, and it is also true that the poor are often unable to care for their own needs, but the argument overlooks the fact that middle-income citizens represent the largest part of the population and contribute the largest total share of revenues to the national budget. They, therefore, should be of concern to the government as well. (Also, is wealth, or the lack of it, the only reason the government should concern itself with certain groups of citizens?)

8. *Hasty Generalization:* Just because one officer does not understand a particular rule does not mean the other people in the administration do not understand it. It is possible that this officer is just not familiar with the rule, or not very bright.

9. *Bifurcation:* Another possibility overlooked here is that although she may not have been aware of what was going on, this does not necessarily mean that she is a fool, for the evidence may have been cleverly hidden from her or was simply not very easy to get at.

11. *Sweeping Generalization:* The rule or prohibition against working on the Sabbath, Jesus was trying to point out to his

accusers, was not designed to be applied to a case where someone was engaged in healing or helping another. Nor, of course, would he have agreed that it was correct to describe it as "working."

15. *Hasty Generalization:* The argument here is: Henry Ford and John D. Rockefeller, like you, were born poor and on a farm, yet they managed to achieve great wealth and fame; the same, therefore, can happen to you. But just because two poor boys succeeded "by honesty and industry" in making it big does not mean that this is all that is needed for that to happen to others. As we know only too well, the American dream is not achieved that simply.

17. *Begging the Question:* Since the phrase *hurried activity* is merely another way of saying "haste," and *careless activity* another way of saying "waste" (since it must often be redone in order to be done correctly), the argument merely repeats itself, reasserting the conclusion in its premise instead of offering proof for it. As a result, no acceptable proof is offered that haste is waste. (After all, some activities do require quickness in their performance.)

18. *Special Pleading:* Mr. Norris sees his son as exhibiting "executive ability," not laziness. But if the situation were reversed and Mr. Norris's son were being persuaded to do Stewart's work, would Mr. Norris see young Stewart as having the same executive ability? Probably not.

19. *Complex Question:* The question assumes that one of the members of the group being addressed left the door open when someone else may have done so.

20. *Question-Begging Epithets:* Until it has been proved that the defendants are truly "hoodlums," it is prejudging the case to label them so. Nor is the use of flattery ("a person of your culture and upbringing") very helpful here.

21. *Begging the Question:* Indeed if matter had always existed then the world could not have been created by God, but that is precisely the issue here: Did matter (i.e., the world) always exist, or did God bring it into being out of nothing.

22. *Begging the Question:* Merely to explain one's terms ("louder . . . [not] nearly as much noise") is not to establish their truth. A different bell ringer may be at work on the same bell, putting more energy into it.

23. *Complex Question:* Has it in fact been established that the students in this class *are* more intelligent than others in the university? The question assumes that they are, but this may not be the case.

25. *Question-Begging Epithets:* This is an example of the use of slanted language to affirm what has not yet been proved. Using such terms and phrases as *conspirators, dastardly plot, behind locked doors,* and *heavily curtained windows* is prejudicial in that it has not yet been proved that the defendants were in fact conspiring together in some nefarious "plot," doing so in great secrecy. If people get together in the privacy of an individual's home to discuss, say, the political situation, that does not mean they are plotting the violent overthrow of the government—even if the discussion is highly critical of the government and its policies.

27. *Special Pleading:* The writer in this circular condemns the use of marijuana as an "obvious crutch" but finds nothing wrong in using God as a crutch to help solve and overcome his problems. The one is a "prop," the other not. Other people might argue, however, that his "leaning on Jesus Christ" is just as much a prop as resorting to drugs, and just as escapist, and that his condemning one and approving the other is a matter of applying a double standard.

29. *Complex Question:* The advertisement asks a question that assumes the stereo system offered is the best without offering proof that it indeed is.

30. *Begging the Question:* A good rest would indeed be a sure cure for insomnia, but that is just the problem: How do you get the ability to rest? It is like saying a cure for sickness is to get well—which is to say nothing.

38. *False Analogy:* The two things compared here differ too greatly for the analogy to be meaningful. It is normally possible for a father to take proper care of his family, the problems generally being well understood and within a person's capacity to deal with them. But this is not the case with the problems of the world or even of one's own country. These are a good deal more complex and not all that well understood. No one nation can suppose itself to be able to deal with them all by itself.

39. *Slippery Slope:* The teacher should state why he or she refuses to answer the question asked, not those that might

be asked. If the question is a proper one, it should be answered; if it gives rise to other questions, these will have to be dealt with as they arise.

41. *Irrelevant Thesis:* The issue is not whether playing outside on the street is harmful to children but whether watching TV is. To argue that watching TV keeps children occupied for hours and off the streets where they might be harmed is to divert attention from the main issue.

42. *False Cause:* Just because UFOs were supposedly sighted shortly before the New England blackout does not mean they were its cause (nor that another blackout like the one in the Southwest is therefore imminent), for what causes power failures are not UFOs but mechanical disruptions.

44. *False Analogy:* Heinlein argues here that the human race is analogous to a single organism such as a tree or a bush that must be "pruned" regularly in order to maintain its growth and health. The differences between the two, however, are so great that the comparison is of little value. Human beings, for one thing, have minds and awareness and are capable of self-direction; individual humans therefore do not need or require "pruning" from others. (Also, Heinlein may perhaps be equivocating with the term *pruning:* as applied to people it means "selective breeding" but as applied to plants it means "clipping off branches.")

47. *Irrelevant Thesis:* The writer sets out to answer the question, Why do I love my country? Instead of an answer to this question, what follows is an irrelevant and sentimental speech about the individual's mother, the hardships of life, and the value of work. Rather than answering the question, the writer seems to have avoided it.

54. *Sweeping Generalization:* The generalization here is that "education results in improvement." However, in the case of "higher education," if one has no "ability," the improvement may be so slight that the effort may not be worthwhile.

56. *Begging the Question:* The argument is circular: (A) Frank is a disreputable person because (B) the people he associates with are disreputable. But how do we know that these people are disreputable? We know this because (B) they hang around with (A) someone as disreputable as Frank! The argument therefore has the following structure: A is so because of B, and B is so because of A.

57. *Complex Question:* This question implies that only "nice" people are married (and insinuates that since the person addressed is not married then perhaps he or she is not that nice). But, as we all know, there are many "nice" people who are not married, and many who are who are not "nice."

58. *Irrelevant Thesis:* What is at issue here, and therefore the only thing to be considered, are the effects of marijuana, not alcohol. To maintain that marijuana, unlike alcohol, does not lead to barroom brawls (assuming this is so) is to divert attention from considering the other possible bad effects of using marijuana.

59. *False Cause:* His hair may have turned white after the wedding for any number of reasons—failing health, financial problems, and the like—none of which is directly connected necessarily to his daughter's marriage.

60. *Irrelevant Thesis:* It may be true that to question our involvement would give comfort to our enemies, but this is irrelevant to whether or not we were in fact right. Thus, though questioning the justice of U.S. policy might give comfort to our enemies, that does not prove that the United States had justice on its side. The country may still have been wrong in doing what it did. (As to whether such questioning is unpatriotic, one might reply that, on the contrary, it would be unpatriotic not to try to ensure that our policies and actions conform to what is right and just.)

61. *Question-Begging Epithets:* What evidence is there that there has been a "deterioration of governmental efficiency," that it is of "deplorable" proportions—all this presumably caused by "widespread indifference" on the part of the populace? Until evidence has been provided to confirm all these charges, we should not allow these labels to influence our attitude and judgment.

62. *Hasty Generalization:* This is a one-sided argument in which only part of the evidence is examined. Everything said here on behalf of high retail prices may be true, but for a balanced judgment some of the contrary evidence (e.g., the burden such prices place on consumers living on fixed incomes, the deleterious effect of high prices on inflation, etc.) should also be considered.

63. *Sweeping Generalization:* Certainly we ought to try to come to one another's assistance, but in this particular case unless the accused scout is a powerful swimmer, the

attempt to save a bigger boy might have resulted in tragedy for both. It would have been better for the scout to have tried to throw the drowning boy a rope or to have run for help.

65. *Bifurcation:* The argument presents us with only two alternatives when, in fact, other alternatives exist. We might try, for example, to change it or ourselves.

67. *Begging the Question:* The argument is circular, beginning where it ends and ending where it begins: (A) the world is good because (B) God is good, and (B) God is good because (A) the world is good. Such an argument succeeds in proving nothing.

68. *Hasty Generalization:* This is a one-sided argument in which only the beneficial results of legalized gambling are presented and all of the possible bad effects are ignored. This is a potentially costly and dangerous way to decide a question.

70. *Sweeping Generalization:* The rule or generalization invoked here is that banks do not sell hot dogs and hot dog vendors do not make loans. It is misapplied here since helping a friend out with a small loan of $5 is not quite like being in the banking business.

71. *Hasty Generalization:* The testimonial asserts that since Mr. Feinstein previously offered successful investment advice, this new idea of buying rare banknotes will also be profitable. It tells us nothing of the soundness of this particular banknote scheme. Instead we are supposed to make our judgment on the performance of a diffferent type of investment. The argument overlooks the fact that there is a great difference between paper money and "paper gold" (as stamps are sometimes called). There is, first of all, not as much interest in banknotes as there is in stamps, which is almost a universal hobby. In addition, banknotes may be more easily counterfeited and may therefore not be such an appealing investment to many.

72. *Special Pleading:* The rabbi is accusing the missionary of applying a double standard. The missionary is willing to accept the fact that since Bar Kochba was killed without bringing the redemption he could not have been the Messiah, yet this does not apparently prevent him from believing in the Messiahship of Jesus who, as the Jews could argue, has also not brought about the redemption.

73. *Irrelevant Thesis:* Whether any doctor *could* have resuscitated the baby is irrelevant to the question of whether Waddill *should* have attempted to save the fetus. Waddill's defense is irrelevant for the further reason that the question here is not whether the fetus could have been saved but whether he deliberately hastened death.

78. *Bifurcation:* The choice is either to spend the weekend with mother or not to. Totally ignored is the possibility of spending part of the weekend with mother and part working on the campaign.

79. *False Analogy:* Thoreau compares a man who eats "animal food" (meat) to that of the larvae state of a butterfly and the maggot state of a fly. He contends that such insects have "voracious" and "gluttonous" appetites when in their larval state and eat much less in their "perfect state" (that is, as a butterfly or mature fly); therefore, humans should also eat much less food and abstain from meat. The analogy falls apart, however, because an insect in its larval state eats a lot in order to store food for its metamorphosis into adulthood. Humans are vastly different from insects in that they do not go through such a metamorphosis. Therefore the human diet, which needs to be more constant and stretched out over a whole life span, cannot be compared to the diet of an insect.

86. *Complex Question:* By way of his last, offensive question, the speaker smuggles in the assumption that minorities spend their time drinking wine and collecting welfare checks.

"And so I ask the jury...is that the face of a mass murderer?"

"The Far Side" cartoon by Gary Larson is reprinted courtesy of Chronicle Features, San Francisco.

5

Fallacies of Relevance

Fallacies of relevance are arguments in which the premises, despite appearances, do not bear upon the conclusions drawn in the arguments. These fallacies might well be called fallacies of irrelevance, for all of them introduce some piece of irrelevance that tends to confuse. What unites this last set of fallacies is that in all of them the irrelevance is an attempt to obscure the real issue by stirring up our emotions. Fallacies of relevance derive their persuasive power from the fact that, when feelings run high, almost anything will pass as an argument.

The six fallacies described in this chapter, although far from exhaustive, represent those frequently met with when the object of argument is to trade on emotional features of our nature such as our susceptibility to prejudice, vanity, pity, fear, and the like. In the first fallacy presented here, that of personal attack, it is our tendency to be bigoted which is exploited. In the fallacy of mob appeal, it is our gullibility that is exploited. The fallacy of appeal to pity exploits our feelings of sympathy, while appeal to authority appeals to both our modesty and our vanity. The last two fallacies appeal to our sense of ignorance and of fear, respectively. All accomplish their objects by introducing irrelevant material designed to evoke a particular emotion in order to establish a conclusion not supported by the premises.

Table of Fallacies of Relevance

Fallacy	Definition/Hints	Example/Method
Genetic Fallacy	Attacking a thesis, institution, or idea by condemning its background or origin	"America will never settle down; look at the rabble-rousers who founded it."
Abusive ad Hominem	Attacking the character of the opposing speaker rather than his or her thesis	"He should clearly not be our leader. He has admitted to being a homosexual."
Circumstantial ad Hominem	Attacking the opposing speaker by implying vested interests	"Sure he opposes rent control; he owns two apartment buildings, doesn't he?"
Tu Quoque	Attempting to show that an opponent does not act in accord with his or her thesis	"How can she tell me to exercise more when I know that all she does is sit behind a desk?"
Poisoning the Well	Attempting to preclude discussion by attacking the credibility of an opponent	"This man has lied his way out of far tougher situations than this. Why should we listen to him?"
Mob Appeal	Using emotion-laden terminology to sway people en masse	"I appeal to you as the most downtrodden and abused people on this earth. Rise up and follow me!"
Appeal to Pity	Seeking to persuade not by presenting evidence but by arousing pity	"Please officer, don't give me a ticket. My parents will take the car away from me and my life will be miserable."

Table of Fallacies of Relevance *(Continued)*

Fallacy	Definition/Hints	Example/Method
Appeal to Authority	Seeking to persuade not by giving evidence but merely by citing an authority, in the form of an: (1) appeal to the one, (2) appeal to the many, (3) appeal to the select few, and (4) appeal to tradition	(1) "If you like people, be sure you brush with Colgate. Walt Frazier wouldn't think of brushing with anything else." (2) "Everybody's wearing it." (3) "The Lancia concept. To set you apart from the crowd." (4) "The institution of marriage is as old as human history and thus must be considered sacred."
Appeal to Ignorance	Emphasizing not the evidence for a thesis, but the lack of evidence against it	"There must be extraterrestrial life. No one has proven there isn't." (Note that this type of argument can be used for both sides of an issue. Consider the analogous opposing argument to the one above.)
Appeal to Fear	Seeking to persuade through fear	"Billy, if you don't go to bed this instant, Santa Claus will take note of it."

1. THE FALLACY OF PERSONAL ATTACK

The fallacy of personal attack is an argument that diverts attention away from the question being argued by focusing instead on those arguing it. The Latin name for this fallacy, *ad hominem,** is probably the most widely used by English-speaking people of the many Greek and Latin words that persist in logic.

---

*Argumentum ad hominem means in Latin, literally, "argument to the man." It is also translated as *"against* the man," a form emphasizing the fact that this fallacy shifts the attack away from the question and places it against the person who is making the argument. (In its sense of an argument *to* the man, the *ad hominem* argument has come to stand loosely for all fallacies of relevance that appeal to our emotional natures rather than our powers of reasoning. In this text, however, we shall restrict its use to fallacies of personal attack against a particular individual or group of individuals.)

Fallacies of personal attack can take various forms, depending on the nature of the attack.

Genetic Fallacy

One of the simplest is *genetic fallacy,* a type of argument in which an attempt is made to prove a conclusion false by condemning its source or genesis. Such arguments are fallacious because how an idea originated is irrelevant to its viability. Thus it would be fallacious to argue that, since chemical elements are involved in all life processes, life is therefore nothing more than a chemical process; or that, since the early forms of religion were matters of magic, religion is nothing but magic.

Genetic accounts of an issue may be true, and they may be illuminating as to why the issue has assumed its present form, but they are irrelevant to its merits.

    a) This scholarship aid proposal is calculated to exploit poor students, for it was written by a committee composed only of members of the faculty and administration. No scholarship students were on that committee.

    b) We must take Schopenhauer's famous essay denouncing women with a grain of salt. Any psychiatrist would at once explain this essay by reference to the strained relationship between Schopenhauer and his mother.

The spread of psychoanalysis has tended to promote the appeal to underlying motivations that is found in argument *b*. Through an unfavorable psychological account of how or why the advocate of a certain view came to hold it, one might claim to undermine any argument whatsoever. But although it may be true that one's motives may weaken one's credibility, motives are irrelevant to the credibility of an argument itself. Arguments are sound not because of who proposes them but by virtue of their internal merit. If the premises of an argument prove its conclusion, they do so no matter who happens to formulate the argument. If they do not, the greatest logician cannot make them sound.

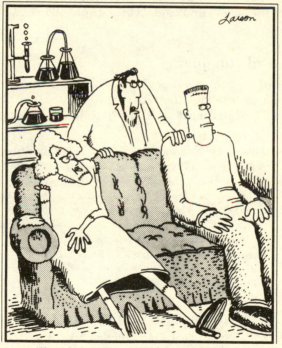

"Hey, c'mon now! . . . You two were MADE for each other!"

"The Far Side" cartoon by Gary Larson is reprinted courtesy of Chronicle Features, San Francisco.

A genetic fallacy. Similar origin might have little bearing on matters of the heart.

If the historical origins of a view are irrelevant to its validity, why do we nevertheless tend to place such emphasis and importance on history? Studying the history of an idea or hypothesis is enormously important in making the idea of hypothesis clearer to ourselves. Seeing how a great mind hit upon a certain idea and going through the same steps he or she took in arriving at it very often helps us grasp it much more easily and understand it better. What such an analysis does not and cannot do is validate or invalidate the idea.

In addition, it often is useful and even sobering to realize that great things and ideas frequently have very humble origins and that the source of the greatness of a work is to be looked for elsewhere. It is useful to remember in this connection the remark made by the sculptor Rodin to the man who served as his model for his famous work, *The Thinker.* After completing it, Rodin turned to the model and said, "O.K., stupid. You can come down now."

Abusive ad Hominem

A variant of the genetic fallacy is the *abusive ad hominem,* which, in addition to drawing attention to the source of an idea, attacks the advocate of that idea with insult or abuse.

    c)   This theory about a new cure for cancer has been introduced by a woman known for her Marxist sympathies. I don't see why we should extend her the courtesy of our attention.

    d)   Oglethorp is now saying that big corporations shouldn't pay more taxes. That's what you'd expect from a congressperson who's lived in Washington for a couple of years and has forgotten all about the people back home.

    e)   In reply to the gentleman's argument, I need only say that two years ago he vigorously defended the very measure he now opposes so adamantly.

Turning attention away from the facts in arguments to the people participating in them is characteristic not only of everyday discussions but of many of our political debates as well. Rather than discuss political issues soberly, rivals may find it easier to discuss personalities and engage in mudslinging. This can be effective because once a suspicion is raised it is difficult to put it to rest. It is not surprising, therefore, that the argument against the individual is all too common in debates among people seeking office.

When feelings run high—or, by contrast, when people simply do not care enough about an issue to give it close scrutiny—abusive tactics such as those above can persuade. Making an opponent appear suspicious, ridiculous, or inconsistent tends to suggest that his or her argument must be unsound because he or she cannot be trusted. Cracking a joke at an opponent's expense, for example, can demolish that person's argument if the joke succeeds in diverting attention away from the issue and causing the opponent to appear silly. A story from the floor of the British Parliament illustrates this well. A member named Thomas Massey-Massey

"I used to be somebody . . . big executive . . . my own company . . . and then one day someone yelled, 'Hey! He's just a big cockroach!'"

"The Far Side" cartoon by Gary Larson is reprinted courtesy of Chronicle Features, San Francisco.

A victim of an abusive ad hominem.

introduced a bill in Parliament to change the name of Christmas to Christ-tide, on the ground that *mass* is a Catholic term that Britons, being largely Protestant, should not use. Another member, it is related, rose to object to the argument. Christmas, he said, might not want its name changed. "How would you like it," he asked Thomas Massey-Massey, "if we changed your name to Thotide Tidey-Tidey?" Amid the laughter that followed, the bill could gain no further hearing.

Circumstantial ad Hominem

Occasionally, instead of engaging in direct abuse, an opponent will try to undercut a position by suggesting that the views being advanced merely serve the advocate's own interests. Logicians call this the "circumstantial" form of the *ad hominem* argument.

Someone making use of the circumstantial form might point out, for example, that a manufacturer's argument in favor of tariff protection should be rejected on the ground that, as a manufacturer, the individual would naturally favor a protective tariff; or that a proposed rent increase must be unjustified because no tenant supports it. Rather than offering reasons why the conclusion in question is true or false, such arguments offer only reasons for expecting that one's opponent might view that conclusion as he or she does.

Although charging an opponent with having vested interests can be seen as a form of reproach, the nonabusive form of this fallacy differs from the abusive form in that abuse is only incidental, not central, to the latter.

> f) It is true that several college professors have testified that these hallucinogenic drugs are harmless and nonaddictive, but these same professors have admitted to taking drugs themselves. We should certainly disregard their views.

Here again, an irrelevancy has been introduced in order to divert attention from the real issue.

We have now observed three of the main tactics used to destroy a person's credibility. It is sometimes done, as we saw, by way of deflation (the genetic fallacy), straight-out insult (the abusive *ad hominem*), or insinuation (the circumstantial *ad hominem*). There are, in addition, two further ways that are somewhat more complicated.

"Personally, I wish the whole world was Jewish."

Reprinted by Special Permission of *Playboy* Magazine: Copyright © 1975 by Playboy.

A witty example of the circumstantial fallacy. The hog's preference is understandable given its own self-interest.

Tu Quoque

The first of these tactics goes by the quaint Latin name *tu quoque,* *
in which the person advocating a position is charged with acting
in a manner that contradicts the position taken. The thrust of the

---

Tu quoque (pronounced "tu kwo-kway") is, in Latin, "you also." In
idiomatic English, it means "look who's talking."

---

tu quoque fallacy is that an opponent's argument is worthless
because the opponent has failed to follow his or her own advice.

> g) Look who's telling me to stop smoking! You smoke more than
> I do.

Although the fact that the suggestion comes from a fellow smoker
tends to weaken its moral force, it does not undermine the argument. The contention that smoking is unhealthy may still be true
whether the person who tells us so smokes or not.

We have a natural tendency to want others to "practice what
they preach." But practice is irrelevant to the merits of an argument. The following retort seems reasonable enough at first
glance but it has no place in logical discourse:

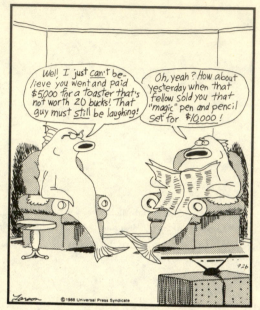

Sucker fish at home

An example of the tu quoque *fallacy.*

h) If you think communal living is such a great idea, why aren't you living in a commune?

Those who resort to this kind of attack often draw courage from another cliché: that people in glass houses should not throw stones. There is no reason, however, why a stone thrown from a glass house cannot find its mark.

It is only a step from an argument charging "You do it too!" to one that charges, "You would do the same thing if you got a chance." Notice this shift between the next two arguments.

i) Far too much fuss has been made over our Central Intelligence Agency's espionage abroad. Other countries are just as deeply engaged in spying as we are.

j) It may be true that Kuwait hasn't yet carried out any espionage activities in the States, but it would if it had the chance. Let's beat Kuwait to it, I say.

The fallacy is fundamentally the same in each case. Whether someone else is already acting in a manner counter to the conclusion at issue—or whether someone else would act in such a manner if the opportunity arose—has no bearing on whether the conclusion in question is right or wrong. As in all fallacies based on personal attack, any consideration of those who hold a position or who originated a position or who are opposed to a position must be viewed as an irrelevance.

Poisoning the Well

The final fallacy of this sort that we will consider is known as *poisoning the well.** In such arguments an attempt is made to place the opponent in a position from which he or she is unable

> *The expression goes back to the Middle Ages, when waves of anti-Jewish prejudice and persecution were common. If a plague struck a community, the people blamed it on the Jews, whom they accused of "poisoning the wells."

to reply. This form of the fallacy received its name from John Henry Cardinal Newman, a nineteenth-century British churchman, in one of his frequent controversies with the clergyman and novelist Charles Kingsley. During the course of their dispute, Kingsley suggested that Newman, as a Roman Catholic priest, did not place the highest value on truth. Newman protested that such an accusation made it impossible for him, or for any other Catholic, to state his case. For how could he prove to Kingsley that he had more regard for truth than for anything else if Kingsley presupposed that he did not? Kingsley had automatically ruled out anything that Newman might offer in defense. Kingsley, in other words, had poisoned the well of discourse, making it impossible for anyone to partake of it.

Consider how these accusations poison the well:

k)   Don't listen to him; he's a scoundrel!

l)   I beg of you, ladies and gentlemen, to remember when you hear members of the opposition that a person opposing this move does not have the welfare of our community at heart.

m) This woman denies being a member of the opposition. But we know that members of the opposition have been brain-

"Hold it, Joey! Give her
a chance!"

"Okay, Margaret . . .
What dumb thing was
you sayin' about Santa
Claus?"

Dennis the Menace® used by permission of Hank Ketcham and © by North American Syndicate.

Poisoning the well is an all-too-common tactic. Is Dennis really giving Margaret a chance for a fair hearing?

        washed to deny under any circumstances that they belong to the opposition.

n) Those who disagree with me when I say that mankind is corrupt prove that they are already corrupted. (Friedrich Nietzsche)

Anyone attempting to rebut these arguments would be hard pressed to do so, for anything he or she said would only seem to strengthen the accusation against the person saying it. The very attempt to reply succeeds only in placing someone in an even more impossible position. It is as if, being accused of talking too much, one cannot argue against the accusation without condemning oneself; the more one talks the more one helps establish the truth of the accusation. And that is perhaps what such unfair tactics are ultimately designed to do: by discrediting in advance the only source from which evidence either for or against a particular position can arise, they seek to avoid opposition by precluding discussion.

Before leaving fallacy of personal attack, it should be pointed out that there are occasions on which it is appropriate to question a person's character. In a court of law, for example, it would not be irrelevant to point out that a witness is a convicted perjurer or a chronic liar. If the assertion is true, this information is relevant. Although this would tend to reduce the credibility of his or her testimony, however, it would not in itself prove that testimony false. Even chronic liars have been known to tell the truth, and we would be guilty of a breach of logic were we to argue that what a person says is a lie because he or she has lied in the past.

2. THE FALLACY OF MOB APPEAL

The *fallacy of mob appeal* is an argument in which an appeal is made to emotions, especially to powerful feelings that can sway people in large crowds. Also called *appeal to the masses,** this fallacy invites people's unthinking acceptance of ideas which are presented in a strong, theatrical manner. Mob appeals are often

---

*Its Latin name is *argumentum ad populum,* literally, "appeal to the people." Like our word *popular,* the term *populum* carries a certain connotation of mass acceptance without thoughtful consideration.

---

said to appeal to our lowest instincts, including violence. The language of such fallacious appeals tends to be strongly biased, making use of many of the linguistic fallacies we have examined previously in this book. Indeed, most instances of mob appeal incorporate other fallacies, melding them together into an argument that rests primarily on appeal to an emotional, rather than a reasoned, response. In so doing, such arguments commit a fallacy of irrelevance because they fail to address the point at issue, choosing instead to steer us toward a conclusion by means of passion rather than reason.

The following example of mob appeal is from the pen of

Westbrook Pegler, long a syndicated columnist of the political right:

> a) Mayor LaGuardia, who himself is a very noisy member of the crowd known as the labor movement, certainly must know that the worst parasites, thieves, and bread robbers now in active practice in the United States, and specifically in the city of New York, are the union racketeers. The waterfront is crawling with them; they are even preying on men employed to produce entertainments for troops under auspices of the USO and thus filching from the fighters for whose benefits the USO funds are raised. Throughout the country, they are reaching their dirty hands into the homes of the poor and stealing bread and shoes from children of the helpless American toiler. LaGuardia has never said a word against such robbery and, by his association with the union movement, he has given approval of this predatory system.

Although this passage reads as if it came about spontaneously, it is the result of careful artifice. Pegler fails throughout to address the point at issue, whether and how the union movement is predatory. Rather, he employs explosive language which is well adapted to lead to the conclusion he wishes us to draw. Mayor LaGuardia is described, for example, as a *noisy,* not just a busy, member of the labor movement. The movement is merely *known as* such; Pegler claims that in reality it is a *crowd* of *racketeers* who are *stealing* from the *helpless American toiler* and his *children.* What kind of people are the unionists? Pegler never uses the word, but his images—*bread robbers, waterfront, crawling with them, preying on, reaching their dirty hands into the homes of the poor and stealing bread, predatory*—leave no doubt that Pegler is comparing the unionists to rats.

At the opposite end of the political spectrum is this passage, from the Communist publication *New Masses* (March 19, 1946), which was directed against British Prime Minister Winston Churchill following his famous "Iron Curtain" speech at Westminster College in Missouri:

> b) Winston Churchill, the archbishop of torydom, came to tell us how we shall live. And what is the life he maps for us? An

Anglo-American tyranny to ride roughshod over the globe. He said that it was against Communism that he wanted the armies and navies combined. The words are Churchill's but the plan is Hitler's. Churchill's own domain of plunder is ripping at the seams and he asks Americans to save it for him. We are to be the trigger men, we are to provide him billions in money to regain what the robber barons are losing.

As in the Pegler passage, suggestive phrases such as *ride roughshod, domain of plunder, trigger men, robber barons* follow one another closely. They are designed to add up to an association of Churchill with a gangster, trying to get us to join his gang. The writer never addresses what it is precisely that Churchill proposes and whether in fact what he proposes can benefit America.

This kind of accusatory writing is not confined to those whose motives may leave something to be desired. Some of our greatest speeches have resorted to it, notably the famous "Cross of Gold" speech delivered to a political convention at the turn of the century by William Jennings Bryan. Bryan's address opposed gold as a monetary standard and favored bimetalism.

c) We care not upon what lines the battle is fought. If they say bimetalism is good, but that we cannot have it until other nations help us, we reply that, instead of having a gold standard because England has, we will restore bimetalism, and then let England have bimetalism because the United States has it. If they dare to come out in the open field and defend the gold standard as a good thing, we will fight them to the uttermost. Having behind us the producing masses of this nation and the world, supported by the commercial interests, the laboring interests and the toilers everywhere, we will answer their demand for a gold standard by saying to them: You shall not press down upon the brow of labor this crown of thorns, you shall not crucify mankind upon a cross of gold.

At the conclusion of this speech, Bryan cast his arms straight out as if nailed to a cross. He then dropped his arms and took one step back. For several seconds there was no sound from the transfixed audience. Then pandemonium broke loose and a tremendous ovation began to swell. Bryan's supporters hoisted the thirty-six-year-old member of Congress to their shoulders and paraded him about the hall. A band struck up a Sousa march. The following day the convention nominated Bryan for president.

We designate mob appeals fallacious because these arguments rest on feelings stirred up in the audience.

> d) I'm a working man myself, and I know how hard it is to make ends meet today.
>
> e) Because you are a college audience, I know I can speak to you about difficult matters seriously.
>
> f) No one in this room wants to deny any child a decent education. But let's remember that this is our school and it belongs to our children.

Those engaging in mob appeal may pretend to fall in with the passions of the crowd they are addressing, or they may flatter them or appeal to their prejudices. Mob appeals invite us to give in to feelings of certitude rather than to rational convictions of certainty. Feelings, however, do not constitute evidence for the truth of a conclusion.

Mark Antony's famous funeral oration over the body of Caesar in William Shakespeare's *Julius Caesar* (act 3, sc. 2) is a brilliant example of mob appeal.* The oration repays study, for the techniques it uses are the stock in trade of propagandists and hate merchants. Antony has been called upon to give the funeral oration because Caesar's assassins believe he will speak sympathetically of the murder. The crowd is addressed first by Brutus, who, fearing that Caesar was preparing to become Rome's dictator, had killed him in a conspiracy with others. Brutus convinces the crowd that he slew Caesar for the good of Rome, and the crowd murmurs that Brutus should be the next Caesar. But Brutus silences them and urges them to hear Antony praise the dead ruler. Antony instead uses the opportunity to incite the mob against Brutus and his fellow conspirators, doing so with such skill that in an incredibly short time he turns the mob into slaves ready to do his bidding.

As Antony makes his way to the pulpit, he hears a citizen exclaim, " 'Twere best he speak no harm of Brutus here"; another cries, "This Caesar was a tyrant." Antony mounts the pulpit, and

---

*I am indebted to Alburey Castell, who in his *A College Logic: An Introduction to the Study of Argument and Proof* (New York: Macmillan, 1946) was the first to make extensive use of this example.

begins with the well-known words, "Friends, Romans, Countrymen"—or as we now would say, "My fellow Americans." Knowing it would be dangerous to try to change their mood abruptly, he tries instead to gain their sympathy by echoing their present feelings. And so he says:

Friends, Romans, Countrymen, lend me your ears.
I come to bury Caesar, not to praise him.
The evil that men do lives after them;
The good is oft interred with their bones.
So let it be with Caesar.

Having won their attention and sympathy, Antony proceeds to sow seeds of doubt regarding Brutus's character and motives. He does so with a display of high intention and modesty.

The noble Brutus
Hath told you Caesar was ambitious.
If it were so [*Notice the qualification.*], it was grievous
   fault,
And grievously hath Caesar answered it.
Here, under leave of Brutus and the rest—
For Brutus is an honorable man;
So are they all, all honorable men—
Come I to speak in Caesar's funeral.
He was my friend, faithful and just to me.
But Brutus says he was ambitious,
And Brutus is an honorable man.
He hath brought many captives home to Rome,
Whose ransoms did the general coffers fill.
Did this in Caesar seem ambitious?
When that the poor have cried, Caesar hath wept—
Ambition should be made of sterner stuff.
Yet Brutus says he was ambitious,
And Brutus is an honorable man.

Antony works the idea of Brutus's honor and Caesar's supposed ambition into every line he can, knowing that if you repeat something often enough people will begin to believe it.

You all did see that on the Lupercal
I thrice presented him a kingly crown,

Which he did thrice refuse. Was this ambition?
Yet Brutus says he was ambitious;
And sure he is an honorable man.
I speak not to disprove what Brutus spoke,
But here I am to speak what I do know.
You all did love him once, not without cause.
What cause withholds you then to mourn for him?
O judgment, thou art fled to brutish beasts,
And men have lost their reason! Bear with me,
My heart is in the coffin there with Caesar,
And I must pause till it come back to me.

Antony's display of grief here, which he says has rendered him
momentarily speechless, marks the turning point of the oration—
away from the question at issue (Brutus's claim that Caesar was
ambitious) and toward an appeal to the mob's emotions.

Pausing to recover his voice, and to let his words sink in, Antony
hears a citizen whisper, "Methinks there is much in his sayings."
Emboldened, Antony begins to plant inflammatory ideas in their
minds.

O Masters! If I were disposed to stir
Your hearts and minds to mutiny and rage [*Of course he
wouldn't be so low as to do that.*]
I should do Brutus wrong, and Cassius wrong,
Who, you all know, are honorable men. [*There he goes
again.*]
I will not do them wrong; I rather choose
To wrong the dead, to wrong myself and you [*Pointing a
finger at them.*]
Than I will wrong such honorable men. [*And again.*]

Still diverting attention from the main issue and inflaming their
passions further, Antony mentions that Caesar has left them a
rich legacy in his will. As expected, at mention of the will, the mob
becomes excited and shouts to have it read. But Antony is not
ready to read it yet because the mob is not yet sufficiently enraged.
Having started a chain of suspense, he goes on to instill further
inflammatory suggestions.

Have patience, gentle friends, I must not read it.
It is not meet you know how Caesar loved you.

You are not wood, you are not stones, but men;
And being men, hearing the will of Caesar,
It will inflame you, it will make you mad: [*There, that
should do it—we can almost hear him saying to
himself.*]
'Tis good you know not that you are his heirs,
For if you should, O, what would come of it!

Knowing full well what would come of it, Antony instructs them
to "make a ring about the corpse of Caesar" so that they may see
"him that made the will." Antony knows that, to be most effective,
the reading requires proper staging. And so, like children, they
gather about the corpse. The crowd that so recently would hear no
criticism of Brutus now stands, subdued and seduced, at Antony's
disposal. They have forgotten Brutus's argument for killing Caesar.

"If you have tears," Antony says quietly, "prepare to shed them
now." He will soften them further by showing them Caesar's
bloody cloak, and he will point out each hole made by an assassin's
dagger. After gaining control of their emotions in this way, Antony
will proceed to demolish Brutus as a traitor.

You all do know this mantle: I remember
The first time ever Caesar put it on.
'Twas on a summer's evening, in his tent,
That day he overcame the Nervii.
Look, in this place ran Cassius' dagger through.
See what a rent the envious Casca made.
Through this the well-beloved Brutus stabb'd,
And as he pluck'd his cursed steel away,
Mark how the blood of Caesar follow'd it,
As rushing out-of-doors, to be resolved
If Brutus so unkindly knock'd, or no.
For Brutus, as you know, was Caesar's angel:
Judge, O you gods, how dearly Caesar loved him!
This was the most unkindest cut of all,
For when the noble Caesar saw him stab,
Ingratitude, more strong than traitors' arms,
Quite vanquish'd him. Then burst his mighty heart;
   [*Caesar died from this shock, not the dagger wounds?*]
And in his mantle muffling up his face,
Even at the base of Pompey's statue,
(Which all the while ran blood) great Caesar fell.

> O, what a fall was there, my countrymen!
> Then I, and you, and all of us fell down,
> Whilst bloody treason [*Openly naming it for the first time.*] flourished over us.

Persuaded that Caesar died not from physical wounds but from pain at Brutus's ingratitude, the mob breaks down and weeps. But Antony has an even more powerful shock tactic in store for them. Throwing off a blanket covering the corpse, he reveals Caesar's mutilated body to the horror-stricken crowd. They become hysterical, shouting: "Revenge! . . . Burn! . . . Kill!"

But Antony holds them back—so that they will want to break loose more than ever and so that he can be sure their fury is not only strong but deep and long-lasting.

> Good friends, sweet friends, let me not stir you up
> To such a sudden flood of mutiny. [*Heaven forbid.*]
> They that have done this deed are honorable, [*Again.*]
> What private griefs they have, alas, I know not,
> That made them do it.

Antony uses here the device of the *big lie,* a lie so big that no one would believe anyone could say such a thing unless it were true. Believing now that it was mere "private griefs" that led Brutus and the others to assassinate Caesar—grievances so trifling that Antony doesn't even know what they were—the crowd becomes enraged. Although they are nearly out of control, Antony still restrains them, reminding them that they have forgotten Caesar's will. He then reads to them the will in which Caesar leaves every Roman citizen a sum of money plus all the royal properties. At its conclusion, Antony says to them, "Here was a Caesar! When comes such another?"

The crowd rushes out, deranged and full of fury. Antony, who is left behind, mutters to himself:

> Now let it work. Mischief, thou art afoot.
> Take thou what course thou wilt!

This magnificent speech helps us see, again, how an argument can be turned away from reason and toward emotion through the cunning introduction of irrelevancies. When the audience is a large group, the enthusiasm stirred up can reach powerful proportions which can bury the real question at issue. Through tactics like sarcasm, suggestion, repetition, the big lie, flattery, and many other devices explored elsewhere in this book, mob appeals exploit our irrationality.

3. THE FALLACY OF APPEAL TO PITY

The *fallacy of appeal to pity* resembles that of mob appeal in that it is an argument designed to win people over to our side by playing on their emotions. It is distinct from mob appeal in that it exploits a single emotion, that of sympathy.*

> *The Latin name of this fallacy is *argumentum ad misericordiam,* literally, "an argument addressed to our sense of mercy."

This fallacy is very common. It is also ancient, as we know from a reference to it in Plato's *Apology,* describing the trial in 399 B.C. of Plato's teacher, Socrates. Speaking to his judges, Socrates says:

a) Perhaps there may be some one who is offended at me, when he calls to mind how he, himself, on a similar or even less serious occasion, prayed and entreated the judges with many tears, and how he produced his children in court, which was a moving spectacle, together with a host of relations and friends; whereas I, who am probably in danger of my life, will do none of these things.

Despite Socrates' stated refusal to employ appeal to pity, he goes on to make explicit use of it.

b) The contrast may occur to his mind, and he may be set against me, and vote in anger because he is displeased at me on this account. Now if there be such a person among you,—mind, I do not say that there is,—to him I may fairly reply: My friend, I am a man, and like other men, a creature of flesh and blood, and not "of wood or stone," as Homer says; and I have a family, yes, and sons, O Athenians, three in number, one almost a man, and two others who are still young; and yet

> I will not bring any of them hither in order to petition you
> for an acquittal. (34C)

Socrates' use of the appeal to pity here is subtle. A more straight-forward example would be this statement to a jury by the re-nowned criminal lawyer Clarence Darrow:

> c) You folks think we city people are all crooked, but we city
> people think you farmers are all crooked. There isn't one of
> you I'd trust in a horse trade, because you'd be sure to skin
> me. But when it comes to having sympathy with a person in
> trouble, I'd sooner trust you folks than city folks, because you
> come to know people better and get to be closer friends.
> (Quoted in Irving Stone, *Clarence Darrow for the Defense: A
> Biography.* New York: Doubleday & Co., 1941, p. 23)

Darrow's argument pleads for sympathy by appealing to the in-nate goodwill of the jurors.

The trouble with such appeals is that, however moving they may be, they may be irrelevant to the issues, in which case they should carry no weight with us. As in all fallacies of relevance, we need to be clear about the question. Thus it would be fallacious for a defense attorney to offer evidence about her client's unfortunate lot as a reason why the court should find him innocent of a crime of which he stands accused. It would not be fallacious, on the other hand, for an attorney to offer such evidence as a reason for treat-ing the accused with leniency.

Like many of the fallacies we have examined, this one too has been exploited by advertisers. An unfortunate example was a full-page advertisement placed by a group of television dealers in 1950, when television was beginning to be mass-produced.

THERE ARE SOME THINGS A SON OR DAUGHTER *WON'T* TELL YOU!

"Aw gee, Pop, why can't we get a television set?" You've heard that. But there's more you won't hear. Do you expect a seven-year-old to find words for the deep loneliness he's feeling?

He may complain—"The kids were mean and wouldn't play with me!" Do you expect him to blurt out the truth—that he's really ashamed to be with the gang—that he feels left out because he doesn't see the television shows they see, know the things they know?

You can tell someone about a bruised finger. How can a little girl describe a bruise deep inside? No, your daughter won't ever tell you the humiliation she's felt in begging those precious hours of television from a neighbor.

You give your child's *body* all the sunshine and fresh air and vitamins you can. How about sunshine for his morale? How about vitamins for his mind? Educators agree—television is all that and more for a growing child.

When television means so much more to a child than entertainment alone, can you deny it to your family any longer?

"Wait! Spare me! . . . I've got a wife, a home, and over a thousand eggs laid in the jelly!"

"The Far Side" cartoon by Gary Larson is reprinted courtesy of Chronicle Features, San Francisco.

An appeal to pity.

Many people found this advertisement in bad taste, and it was soon withdrawn. Present-day advertisements are somewhat more subtle, but irrelevant appeals to sympathy persist, as in Avis Rent-a-Car's successful slogan: "We're number two."

4. THE FALLACY OF APPEAL TO AUTHORITY

We make an appeal to authority whenever we try to justify an idea by citing some source of expertise as a reason for holding that idea. Appeals to authority are often valid, as when we tell someone to use a certain medicine because the doctor has prescribed it. But appeals to authority can be fallacious, as when we cite those who have no special competence regarding the matter at hand. The *fallacy of appeal to authority,* therefore, is an argument that attempts to overawe an opponent into accepting a conclusion by playing on his or her reluctance to challenge famous people, time-honored customs, or widely held beliefs. The fallacy appeals, at base, to our feelings of modesty,* to our sense that others know better than we do.

---

*The British philosopher John Locke gave this fallacy its Latin name, *argumentum ad verecundiam,* literally, "an argument addressed to our sense of modesty." *Verecundiam* carries connotations of shame as well as modesty, emphasizing how we may be browbeaten into accepting an erroneous conclusion because we are ashamed to dispute supposed authority.

The Authority of the One

It seems obvious that a person's competence in physics or business does not automatically make him or her an expert in politics or gardening. Yet examples of appeals to the *authority of the one* abound in everyday discussion as well as in more structured arguments.

a) This idea of supporting research on dolphins is ridiculous. We were discussing it during lunch today, and the treasurer, the president, and the personnel manager all agreed it's a colossal waste.

Unless someone has taken the trouble to become informed about the particulars of dolphin research, his or her opinion on the subject cannot be considered authoritative and must be dismissed as irrelevant to the question. Even experts do not ask that their opinions be accepted because they say so but because those opinions are derived from evidence.

The human race, and science in particular, has sometimes paid a steep price for this reverence for authority that seems ingrained in us. The following argument favoring Ptolemy's theory of the solar system over that of Copernicus is one among many examples to be found in the history of science:

b)   One may doubt whether it would be preferable to follow Ptolemy or Copernicus. For both are in agreement with the ob-

"You know, Russell, you're a great torturer. I mean, you can make a man scream for mercy in nothing flat ... but boy, you sure can't make a good cup of coffee."

Authority in one domain does not imply authority in unrelated areas.

served phenomena. But Copernicus's principles contain a great many assertions which are absurd. He assumed, for instance, that the earth is moving with a triple motion, which I cannot understand. For according to the philosophers a simple body like the earth can have only a simple motion. . . . Therefore it seems to me that Ptolemy's geocentric doctrine must be preferred to Copernicus's doctrine. (Clavius, 1581)

But the Copernican theory survived, while Ptolemy's did not. Phrases such as *according to the philosophers* appeared often in the writings of even the best scientists during the Renaissance, bearing witness to the struggle science had to wage against "authorities." A letter of this period, from the Italian astronomer Galileo to his German colleague Johannes Kepler, suggests the extent of the problem. Galileo had just made use of the telescope for the first time, but the authorities refused to look through it.

c) Oh, my dear Kepler, how I wish that we could have one hearty laugh together! Here, at Padua, is the principal professor of philosophy whom I have repeatedly and urgently requested to look at the moon and planets through my glass, which he pertinaciously refuses to do. Why are you not here? What shouts of laughter we should have at this glorious folly? And to hear the professor of philosophy at Pisa, laboring before the grand duke with logical arguments, as if with magical incantations, to charm the new planets out of the sky. (Quoted in Elma Ehrlich Levinger, *Galileo: First Observer of Marvelous Things.* New York: Julian Messner, 1961, p. 84)

In the face of such stubborn adherence to tradition, it is not surprising that some scientists and institutions came to adopt as their credo: "Authority Means Nothing."

If modern science has liberated itself to a great extent from irrelevant appeals to authority, some aspects of our society still cling to the notion of expertise. News reports are sprinkled with phrases such as "Official sources hinted . . ." or "An unidentified spokesperson disclosed . . ." without any indication whether the source is a head of state or a fellow reporter in the next phone booth.

The Authority of the Many

Contemporary America seems especially prone to the form of the fallacy of appeal to authority known as *argument by consensus* or appeal to the *authority of the many*. An out-and-out appeal to numbers, this form of the fallacy is widely used in advertising, where the fact that millions of Americans use a certain product is advanced as a reason for buying it.

> d) In Philadelphia nearly everybody reads the *Bulletin*.
> e) Sony. Ask Anyone.
> f) Last year, over 5,000,000 cats switched to Tender Vittles.

Such appeals are like saying that a book must be good because it is a best-seller or that a film that people are lining up to see all over the country must be a great film. The book or film may be successful, but success is irrelevant to the question of merit.

When former Senator George Smathers of Florida made the following accusation, he invoked so many authorities that his argument seems irrefutable at first glance:

> g) I join two presidents, twenty-seven senators, and eighty-three congressmen in describing Drew Pearson as an unmitigated liar.

As in all cases of this fallacy, however, the fact that many people agree with a certain conclusion does not make it true.

Although the consensus form of the fallacy of appeal to authority is prominent today, it has a long past. At the beginning of this century, the Italian philosopher Benedetto Croce analyzed the horrors of the Spanish Inquisition according to a consensus argument.

> h) The Inquisition must have been justified and beneficial, if whole peoples invoked and defended it, if men of the loftiest souls founded and created it severally and impartially, and its very adversaries applied it on their own account, pyre answering to pyre. (*The Philosophy of the Practical*, trans. Douglas Ainslie. London: Macmillan, 1913, pp. 69–70)

Clearly, the fact that many people endorsed the Inquisition did not make it right. Much older is this argument from an early British chronicle:

"Okay, Williams, we'll vote . . . how many here say the heart has four chambers?"

"The Far Side" cartoon by Gary Larson is reprinted courtesy of Chronicle Features, San Francisco.

An appeal to the many, or authority by consensus.

> i) If the proposal be sound, would the Saxon have passed it by? Would the Dane have ignored it? Would it have escaped the wisdom of the Norman?

The chronicler's argument is erroneous in suggesting that whatever the other illustrious races may have done or not done has any bearing on the merits of the proposal before him.

The Authority of the Select Few

Just as we should guard against being taken in by an appeal to the authority of a single expert, or of the many, we must also be able to recognize appeals to the *authority of the select few.* Sometimes called *snob appeal,* this form of the fallacy of appeal to authority exploits our feeling that we are aristocrats at heart, that we belong

not to the mass but to the select few. The use of glamorous personalities to advertise products trades on snob appeal, as do advertisements such as the following:

j)   Camel Filters. They're not for everybody.
k)   We make the most expensive wine in America.
l)   Only one grape in 50 grows up to become great champagne.
m)   Guerlain is pleased to announce that only one man in ten
     thousand wears Imperiale.
n)   Adventure Prints by Armstrong. If you've got the courage,
     we've got the carpet!
o)   The lady has taste. Eve. Regular and menthol.
p)   Introducing the Date Walnut Double Decker. It's not your ev-
     eryday cookie.

The authority appealed to in such arguments is that of prestige or
exclusivity. These qualities would not be irrelevant if the arguments had set out to prove that the products in question were
prestigious, but the object of such advertisements is to lead to the
conclusion that we ought to buy those products.

The Authority of Tradition

The appeal to authority can take still another form, an appeal to
custom or tradition, as in this argument:

q)   The institution of marriage is as old as human history and thus
     must be considered sacred.

To this we might reply that so is prostitution. Is it to be regarded
as sacred too?
   The appeal to custom or tradition has not been overlooked by
advertisers. For example, one liqueur advertisement carries the
headline:

r)   Drink Irish Mist, Ireland's legendary liqueur.

Essentially, that is the same appeal as the one made by advertisers
who tout their products as "old-fashioned" and "like grandma
used to make."

5. THE FALLACY OF APPEAL TO IGNORANCE

The *fallacy of appeal to ignorance** is an argument that uses an opponent's inability to disprove a conclusion as proof of the conclusion's correctness. By shifting the burden of proof outside the

---

*The Latin name of this fallacy is *argumentum ad ignorantiam.*

---

argument onto the person hearing the argument, such an argument becomes irrelevant. One's inability to disprove a conclusion cannot by itself be regarded as proof that the conclusion is true.

The following two arguments attempt to shift the burden of proof:

  a) There is intelligent life in outer space, for no one has been able to prove that there isn't.
  b) I know that every action we perform is predetermined because no one has proved that we have free will.

Such fallacious arguments involve an appeal to the emotions in that one hopes to place opponents on the defensive, causing them to believe that the proposed conclusion must be true merely because they cannot prove otherwise. That belief would be irrational, resulting from the feeling of intimidation. In logical argument, it is always the obligation of those who propose conclusions to provide the proof.

If the absence of evidence against a claim could be counted as proof for it, we could prove anything we liked—including miracles, as the medieval legend of the weeping statue illustrates. On Good Friday of each year while the congregation bowed in prayer, it was said, the statue on the church altar would kneel and shed tears. But if even one member of the congregation looked up from prayers in order to see the tears, the miracle would not occur. The statue would cry only when all members of the congregation exhibited complete faith.

The legend seems a pleasant, harmless parable, but the fallacy of appeal to ignorance can lead to more problematic conclusions.

  c) The chiropractors have failed entirely in their attempts to establish a scientific basis for their concepts. This question can

> therefore be settled once and for all. Chiropractic has no
> basis in science.

The fact that chiropractors have failed so far to prove that their
concepts have a scientific base is no proof that they lack such a
base, which may yet be established.

> d) There is no proof that the dean leaked the news to the papers,
> so I'm sure she couldn't have done such a thing.

Although we are not at present in a position to show that she did
it, this constitutes no proof that the dean is innocent.

It should be noted, however, that this mode of argument is not
fallacious in a court of law, where, if the defense can legitimately
claim that the prosecutor has not established guilt, then a verdict
of not guilty is warranted. Although this claim may seem to commit the fallacy of appeal to ignorance, it does not really do so. The
relevant legal principle in this case is that a person is presumed
innocent until proved guilty. We do not say that defendants *are*
innocent until proved guilty, but only that they are presumed to be
so. Clearly, not every person whose guilt has not been proven is
innocent; some are later proved, in fact, to be guilty. But until this
is done, they are regarded as legally innocent, whether they are
actually innocent or not. Practice in court is therefore not really
an exception to our rule.

Someone making use of an appeal to ignorance will frequently
try to strengthen a case with question-begging epithets.

> e) I have never once heard an argument for price controls that
> any sensible person would accept. Therefore price controls
> are obviously a bad idea.

This argument is objectionable not only because it bases its proof
on ignorance but also because its use of the epithet *sensible* condemns out of hand any arguments for price control that we might
be ready to offer. Notice the question-begging epithets in these
fallacies:

> f) No responsible scientist has proved that the strontium 90 in
> nuclear fallout causes leukemia. Therefore we can disregard

the alarmists and continue testing nuclear weapons with a clear conscience.

g) If there were any real evidence for these so-called flying saucers, it would be reported in our reputable scientific journals. No such report has been made; therefore there is no real evidence for them.

The terms *responsible scientist* and *alarmist* in argument *f* and *real* and *reputable* in argument *g* render us virtually incapable of responding without being labeled as alarmist or disreputable. Anyone who states an argument in this form of the fallacy does not want to argue logically and is not prepared to be contradicted on logical—or any other—grounds.

In studies of the appeal to ignorance, it is sometimes pointed out that the language of some marriage ceremonies seems to commit this fallacy, when the person officiating asks whether anyone present can show just cause why the bride and groom should not be married. Yet, in stating that anyone who can show such cause should now come forward or forever after be silent, no argument is involved but only a simple statement. It does not attempt to prove that, if no one can show cause why the marriage should not take place, then no such cause exists. It says only that, if no one objects to the marriage, then the marriage will proceed.

6. THE FALLACY OF APPEAL TO FEAR

The *fallacy of appeal to fear* is an argument that uses the threat of harm to advance one's conclusion. It is an argument that people and nations fall back on when they are not interested in advancing relevant reasons for their positions. Also known as *swinging the big stick,** this argument seldom resolves a dispute.

---

*As the Latin word for stick or staff is *baculum,* this argument is known in Latin as *argumentum ad baculum.*

This argument should be distinguished from an all-out threat. If someone should hold a gun to your back and say, "Your money or your life," it would not do to reply, "Ah ha! That's a fallacy!" It is not a fallacy because it is not an argument. Although the gunman is appealing to your sense of fear, and even offering a reason why you should do what he tells you, he is not offering evidence in support of the truth of some statement. He is not arguing with you; he is simply ordering you. The same would be true of threats such as "If you don't keep that dog out of my yard, I'll shoot it," or "You get back into bed or I'll spank you."

Compare the above examples with the following one, in which an attorney says to a jury:

> a) If you do not convict this murderer, one of you may be her next victim.

This is an argument, although a fallacious one. A reason is being advanced for an action being advocated, namely conviction. But the issue of fear that is raised is irrelevant to the real issue, which is not conviction but guilt. Expressed fully, this argument is as follows:

> b) You must believe with me that this woman is guilty of the crime of which she is accused, for if you do not find her guilty of it she will be released and you may end up being her next victim.

Put this way, the argument is clearly fallacious, for what a defendant might do in the future is not relevant to—and in no way proves either his or her innocence or guilt of—a crime committed in the past. It is the fear stirred up by the attorney which deceives us into thinking the argument is relevant.

An appeal to fear therefore offers fallacious evidence. In some cases the evidence will be brief and implicit, as in examples *a* and *b* above; in other cases it may run to pages or even volumes. The following account of an Athenian's appeal to representatives of the island of Melos to join them, taken from the Greek historian Thucydides, is a classic example of an appeal to force.

> c) You know as well as we do, that, in the logic of human nature, right only comes into question where there is a balance of power, while it is might that determines what the strong exhort and the weak concede.... Your strongest weapons are hopes yet unrealized, while the weapons in your hands are

somewhat inadequate for holding out against the forces already arranged against you.... Reflect... that you are taking a decision for your country ..., a country whose fate hangs upon a single decision right or wrong. (Adapted from *History of the Peloponnesian War,* bk. 5, ch. 7)

Not all threats are expressed as elegantly as this. We may encounter the appeal to fear in language like the following:

d) Don't argue with me. Remember who pays your salary.
e) You don't want to be a social outcast, do you? Then you'd better join us tomorrow.
f) This university does not need a teacher's union, and faculty members who think it does will discover their error at the next tenure review.

These arguments are crude forms of the fallacy. They are explicit about the threats being issued.

The fallacy also lends itself to veiled threats.

g) Dear Editor, I hope you will agree that this little escapade by my son has no real news value. I know you'll agree that my firm buys thousands of dollars worth of advertising space in your paper every year.
h) Refugees who come to this country from Southeast Asia have no right to try to preserve their own customs and language while rejecting American ways. If it weren't for fine American customs, those people would go to the dogs.

Although these arguments do not state a threat explicitly, there can be little question of the nature of the reasons being advanced.

Appeals to fear tend to multiply during periods of stress or conflict, both among nations and among individuals. When the oil-producing countries of the Middle East assumed a position of threat to the oil supplies of the West in 1974, for example, it became known that they had compiled a list of U.S. companies that they judged to favor Israel. This was interpreted as an appeal to fear directed at American firms that might favor Israel in the Arab-Israeli dispute. Similarly, during the period of strongly anti-Communist activities of the John Birch Society in this country, the

"Uh-oh, Donny. Sounds like the monster in the
basement has heard you crying again. ... Let's
be reaaaal quiet and hope he goes away."

*Children, and people who are ill-informed, are most susceptible
to appeals to fear.*

society issued the following statement, preceding a list of names
of advertisers:

i)   This list of advertisers has been compiled so that members of
the Anti-Communist Conservative Movement will have a
way of determining those firms that are standing up to be
counted in the fight to preserve our American heritage.

Although the statement went on to point out that there was no
suggestion intended that those whose names did not appear were
therefore Communists, an implication to that effect was clear nevertheless. For in fallacies of irrelevance, to say what one does not
intend to say may amount to the same thing as saying it in the first

place. As in all fallacies of irrelevance, the object of the argument is an appeal to emotion rather than to reason.

SUMMARY

This chapter has presented six *fallacies of relevance*. Such fallacies were shown to be arguments in which the emotional appeal deceives us into believing that what is said is relevant to the conclusion being urged when the real object of the appeal is to enlist support for the conclusion through an emotional rather than a logical response.

Personal attack was shown to be an argument which attacks the person or persons associated with a question rather than attacking the question itself. An example was the argument dismissing a cure for cancer because its discoverer was known for her Marxist sympathies. We saw that this fallacy, which plays on suspicions and prejudices, can take several different forms: genetic fallacy where one casts aspersions on the source of the argument; the abusive and circumstantial ad hominems where one resorts to insult and innuendo; the *tu quoque* and poisoning the well where one raises doubts about an opponent's consistency and credibility.

Mob appeal was seen to result from various propagandist techniques used in combination to arouse people's emotions and to divert attention away from the real question. Mark Antony's funeral oration was explored as a striking example.

Appeal to pity was shown to result from trying to gain one's point by playing on people's sympathies. The advertisement beginning, "There are some things a son or daughter *won't* tell you!" was cited as an illustration.

Appeal to authority was seen as an attempt to intimidate us into accepting a conclusion because someone who is presented as an expert has endorsed it, although that person proves on examination not to be an expert on this subject; because large numbers of people have endorsed it; or because it is endorsed by the glamorous among us; or by tradition. The slogan, "In Philadelphia nearly everybody reads the *Bulletin*" was one example given.

Appeal to ignorance was shown to be an argument that tries to intimidate us into believing that something must be so merely because we cannot prove it is not so. If, for example, we cannot prove that the dean leaked the story to the papers, we must assume that she did not do so.

Appeal to fear, finally, was seen as the fallacy that arises when support for a conclusion is enlisted not by establishing that conclusion through logical inference but by the use of a threat of force or harm. The argument that "if you do not convict this murderer, one of you may be her next victim" was shown as exemplifying this fallacy.

EXERCISES

Identify the fallacy of relevance—personal attack, mob appeal, appeal to pity, appeal to authority, appeal to ignorance, or appeal to fear—that is committed in or that would result from each of the following. Explain the error committed in each case.

* 1. If your idea were any good, someone would have thought of it already.

2. The most recent occurrence of recent years is all these knuckleheads running around protesting nuclear power—all these stupid people who do not research at all and who go out and march, pretending they care about the human race, and then go off in their automobiles and kill one another. (Ray Bradbury, in *Omni*, October 1979)

3. A study of primitive tribes shows that early man had many fears—his principal fears were those of illness, being crushed under a falling tree, and being killed by wild animals. Certain men gained influence in tribes by offering various charms and incantations to ward off these dangers or by asserting that some benevolent spirit more powerful than these dangers would protect those who approached him in the right way. The ones who preferred the second policy were those who introduced religion. Since this is the origin of belief in God, it is little more than superstition. (Anthropology text)

* 4. No, if you don't mind losing a tire, going off the road, and maybe killing yourself, you don't need a new tire.

5. Old soldiers never die; they just fade away.

* 6. I suppose *you* never did anything wrong?

* 7. No breath of scandal has ever touched the senator. There-fore he must be incorruptibly honest.

* 8. Congress shouldn't bother to consult the Joint Chiefs of Staff about military appropriations. As members of the armed forces, they will naturally want as much money for military purposes as they think they can get.

9. It must be so. I read it in a psychology book.

10. You are in a strange town, strolling aimlessly down a street. You could—if you don't fully understand your rights—wind up in jail for the night. (Advertisement for *Time-Life Family Legal Guide*)

*11. A basic weakness of the work, however, is that the hearer cannot help identifying some of the emanations from the [stereo] speaker with more mundane associations. You hear roars from an air terminal, background effects from a cheap radio thriller, the staccato click of rolling dice, screeching brakes, or the unpleasant vibrato of an electronic organ. . . . What value does such a work have as music? Your guess is as good as mine. . . . One thing I do know. Luening and Ussachevsky cannot be dismissed as playboy pranksters. Too many men, not only in this country but also on the continent, are following their same line of thought. (Critic William Mootz, on the *Rhapsodic Variations for Tape Recorder and Orchestra* by Luening and Ussachevsky)

12. All loyal Americans will deplore the passage of this bill.

13. There's no point in listening to what you have to say; everybody knows you're on the radical right.

14. You'd better find the fallacies in these arguments and identify them correctly, for if you don't I'll know you haven't been paying attention in class.

*15. As expected the opposition to Proposition 13 [to reduce state property tax] is signed by two persons long on the taxpayers' payroll and one person from a tax-free foundation. These people are worried about their jobs, not the public sector. (Letter to the editor)

16. We favor extended social security. Isn't that the American thing to do? And we are trying to reduce social discrimination. That's the American way too.

Auth[ority] 17. The golden rule is basic to every system of ethics ever devised. Everyone accepts it in some form or other. It is, therefore, an undeniably sound moral principle.

Mob 18. Let's win this one for Ike. (Richard Nixon's 1968 acceptance speech at the Republican National Convention)

* 19. How should a man avoid pessimism who has lived almost all his life in a boarding house? And who has abandoned his only child to illegitimate anonymity? At the bottom of Schopenhauer's unhappiness was his rejection of the normal life—his rejection of women and marriage and children. (Will Durant, *The Story of Philosophy*)

Pity 20. I'm on probation, Sir. If I don't get a good grade in this course, I won't be able to stay in school. Please, could you let me have at least a C?

Fear 21. Gentlemen, I am sure that if you think it over you will see that my suggestion has real merit. You should look upon it as only a suggestion, of course, and not an order, even though I am the chairman of the board.

* 22. The Cadillac Eldorado. Life is too short to put it off for long.

Arg[umentum] 23. It's the old time religion, and it's good enough for me.

24. A substantial federal subsidy for the lettuce growers is
Pity sorely needed. These growers have suffered from a late spring, a killing frost, and a serious shortage of field help. The resulting loss has depressed the entire Imperial Valley retail industry.

25. It's clear that Jones is an innocent man. He was tried for the crime before a jury of his peers and the prosecution was unable to prove him guilty.

26. The inconveniences of the gas shortage are bad enough. But what's worse is what could happen to you on the road. Because that's where the *real* crisis waits.

In a Volvo, good gas mileage is standard. But it's not the sole attraction. Volvos come with much more important things. Like superior braking, handling, performance and construction which protect you and your passengers. And give you superior value for your money.

So before you buy any car merely because it claims to get good mileage, stop in at a Volvo showroom. There's a lot more you could save.

VOLVO. A car you can believe in. (Advertisement)

27. I don't care how sick she is. She is wanted at the shop immediately. When the supervisor sends for someone, the employee is expected to show up.

*28. A vote for my opponent is a vote for war.

29. DEAR ABBY:

   I must comment on the controversy concerning bullfighting. What surprises us Spanish people is that Americans condemn bullfighting but see nothing wrong with the outrageous savagery of their so-called sport, boxing. How can you condemn bullfighting and approve the brutality of two human beings beating each other with the ultimate purpose of knocking the opponent unconscious, which at times has caused fatal results? It is inconceivable to us that a civilized country permits such inhuman contests causing swollen eyes, bloody noses, and faces beaten to a bloody pulp! Abby, please be less concerned about our bulls, and show more compassion for your fellow man by trying to outlaw boxing.

*30. The Canadian government is having similar thoughts [about banning seal hunting] after four years of hostile publicity and occasional exaggerations about the hunt. Angry letters and petitions flood Ottawa, and demonstrations have besieged Canadian embassies and consulates. Among the protestors are Americans obviously unaware that the U.S. sanctions hunters who annually club or shoot 120,000 seals in the Pribiloff Islands off Alaska. (*Time,* March 21, 1969)

31. The God that holds you over the pit of hell, much as one holds a spider or some loathsome insect over the fire, abhors you, and is dreadfully provoked; his wrath towards you burns like fire; he looks upon you as worthy of nothing else, but to be cast into the fire; you are ten thousand times so abominable in his eyes as the most hateful and venomous serpent is in ours. You have offended him infinitely more than a stubborn rebel did his prince; and yet it is nothing but his hand that holds you from falling into the fire every moment. (Jonathan Edwards, "The It of Hell," 1741)

32. BLACK: Well anyway, when we're silent nobody is playing games.
   WHITE: Silence itself may be a game.
   RED: Nobody was playing games today.

WHITE: But not playing games may itself be a game.
(Eric Berne, *Games People Play*)

* 33. The phenomena reported by the medium are so unusual that extremely good evidence is needed before we can believe them. And it is impossible to obtain the required evidence because we cannot trust people who believe in such phenomena.

34. But let us set aside the dry legalistic aspects of this issue and get to the simple heart of the matter, which can be put in a nutshell. Picture, if you will, the young lad sent off to service abroad, by the vote of his senator or congressional representative. He has nothing to say about it; he is pushed around by his so-called representative, but he cannot vote; he can only silently grieve.

* 35. The hullabaloo over dishonesty among athletes at the All-State Conference is unfounded, for the mayor who attended that conference declared unequivocally that he saw no evidence of dishonesty whatsoever.

36. Let's get down to brass tacks. None of us here is a Philadelphia lawyer; we're just plain folks, trying to see our way clear. There's been a lot of high-falutin' talk about "economic implications" and such like, but the plain fact is that if they build that dam here it will cost us money we just don't have. I'm against it—we're all against it.

* 37. Bishop Wilberforce scored a telling hit in his famous debate with Thomas Huxley on the subject of evolution. He simply inquired casually whether Huxley was descended from the monkey on his mother's side or his father's side of the family.

* 38. Socrates, I think that you are too ready to speak evil of men: and, if you will take my advice, I would recommend you to be careful. Perhaps there is no city in which it is not easier to do men harm than to do them good, and this is certainly the case at Athens, as I believe that you know. (Anytus in Plato's *Meno,* 94E)

39. In William Shakespeare's *The Merchant of Venice* (act 4, sc. 1), the Duke tells Shylock that he should be merciful and not demand by law that he get his bond, a pound of Antonio's flesh. To this Shylock replies:

You have among you many a purchased slave,
Which, like your asses and your dogs and mules,

You use in abject and in slavish parts.
Because you bought them: shall I say to you,
Let them be free, marry them to your heirs?
Why sweat them under burdens? let their beds
Be made as soft as yours and let their palates
Be season'd with such viands? You will answer,
"The slaves are ours": so do I answer you:
The pound of flesh, which I demand of him,
Is dearly bought: 'tis mine, and I will have it.

40. Speculation, revelations, and reflections three days before
the Los Angeles Rams begin the annual summer dullness
known as preseason games: Will George Allen start Isiah
Robertson Saturday evening against New England? The
Rams coach won't say but it's known several of his players
have informed Allen through an emissary that "serious
morale problems" will ensue if Robertson gets the nod over
Bob Brudzinski. "Brudzinski better start," warned one
player. (News item)

41. "I'm a Pepper, he's a Pepper, she's a Pepper, we are Pep-
pers, wouldn't you like to be a Pepper, too?" (Dr. Pepper
soft drink commercial)

42. The Soviet Union has not changed since Stalin's time. It
has one course and one course only. It is dedicated to the
belief that it is going to take over the world. Moreover, the
Soviets have been winning everywhere for twenty-five
years because of a U.S. foreign policy bordering on ap-
peasement. Washington has seriously weakened U.S. de-
fenses, and what is needed is a rapid buildup in all types
of arms. Tune out those cynics, pacifists, and appeasers
who tell us the army and navy of this country are nothing
but extensions of some malevolent military-industrial
complex. There is only one military-industrial complex
whose operations should concern us, and it is not located
in Arlington, Virginia, but in Moscow. (Ronald Reagan, in
his presidential campaign, 1980)

43. After a long trial, the jury let Bert Lance go. Lance denied
the charges, as well as any intention of hurting the banks.
In an emotional summation that left some jurors weeping,
Defense Attorney Nickolas P. Chilivis told them: "Those
folks in Washington can't understand how we trust folks
down here. If you find Mr. Lance guilty of anything, you

will have ruined the reputation, life and character of one of the South's finest men." (News report)

44. "Anyone who knows anything about recent Chinese history," said the professor, "would recognize at once that what has happened since Mao tse-Tung's death is really only the latest episode in over a century of efforts by the Chinese to modernize their economy in response to the impact of the Western nations on China."

45. There was only one catch and that was Catch-22, which specified that a concern for one's own safety in the face of dangers that were real and immediate was the process of a rational mind. Orr was crazy and could be grounded. All he had to do was ask; and as soon as he did, he would no longer be crazy and would have to fly more missions. Orr would be crazy to fly more missions and sane if he didn't, but if he was sane he had to fly them. If he flew them he was crazy and didn't have to; but if he didn't want to he was sane and had to. (Joseph Heller, *Catch-22*)

46. The unique Celebrity hairpiece is the choice of all the "Beautiful People" from motion picture and TV stars to international celebrities. All prefer the remarkable realism of a Celebrity to any other hairpiece or replacement method known. Custom designed and exclusively created only by Bob Roberts, formerly director of the Max Factor Hair Division. Discover the amazing difference for yourself! (Advertisement)

47. Billy Martin, after resigning as Yankee manager, saying why he would make no further comment: "That means now and forever, because I am a Yankee and Yankees do not throw rocks."

48. The moment Alice appeared, she was appealed to by all three to settle the question, and they repeated their arguments to her, though, as they all spoke at once, she found it very hard to make out exactly what they said.

The executioner's argument was, that you couldn't cut off a head unless there was a body to cut it off from: that he had never had to do such a thing before, and he wasn't going to begin at *his* time of life.

The King's argument was, that anything that had a head could be beheaded, and that you weren't to talk nonsense.

The Queen's argument was that, if something wasn't done about it in less than no time, she'd have everybody executed, all round. (It was this last remark that had made the whole party look so grave and anxious.) (Lewis Carroll, *Alice in Wonderland*)

*49. Great American Soup. About as close as you can get to homemade without making it yourself. (Advertisement)

50. The apple of our eye. Is there anything that can top Mom's apple pie? Our natural cheddar. Cracker Barrel cheddar cheese. It's another reason America spells cheese K-R-A-F-T. Our pride. Your joy. (Advertisement)

Find examples from your daily reading of the fallacies discussed in this chapter and explain in detail why you think they are fallacious.

ANSWERS TO STARRED EXERCISES

1. *Personal Attack:* This abusive *ad hominem* argument is an insulting remark, intimating that the person is not bright enough to have been the first to think of the idea. Someone, however, had to be first to do so, and it could well have been that person.

4. *Appeal to Fear:* This is an attempt to frighten the person addressed into buying a new tire or tires. Not a word is said about the condition of the present tires and whether they need to be replaced; rather, only a list of the dreadful things that can happen to motorists is supplied.

6. *Personal Attack:* This is a *tu quoque* argument. Two wrongs do not make a right. The fact that another person too is often blameworthy does not make the speaker less so. Both may be blameworthy.

7. *Appeal to Ignorance:* The fact that no "breath of scandal" has ever touched the senator is no proof that he is "incorruptibly honest." He may simply have been very lucky or clever in evading discovery. Needless to say, that there is no proof that he is "incorruptibly honest" does not mean that he is not. We would be guilty of the same fallacy if we thought so.

8. *Personal Attack:* This is the circumstantial *ad hominem* form of the fallacy. Whether the appropriations requested are justified can only be decided by looking at the requests and not the presumed motives of those making them. Cer-

tainly the Joint Chiefs of Staff are interested parties—and they should be—but this does not necessarily make them incapable of impartial judgment. To think so is to cut ourselves off here from one of the main sources of information needed to make a sound judgment.

11. *Appeal to Authority:* The fact that the music in question has many followers ("too many men, not only in this country but also on the continent") is in itself no proof of its worthiness. They could all be mistaken. At one time everyone thought the earth was flat.

15. *Personal Attack:* Rather than examining the reasons that the persons mentioned in the letter may have for opposing the bill to reduce state property taxes, the writer of the letter accuses them, by way of the circumstantial fallacy, of vested interests: they are opposed to the bill, the writer says, because if it goes through, the state will have less money and their jobs may be eliminated.

19. *Personal Attack:* Durant is guilty of the genetic form of the fallacy. Whether Schopenhauer's views concerning pessimism are sound can only be determined by examining the views themselves and the arguments he brings to bear in support of them. How Schopenhauer came to adopt these views, their genesis, is irrelevant to their possible truth or soundness. Many people have arrived at their views and theories in strange ways which are interesting to learn about but do not and cannot affect the correctness or incorrectness of the views or theories.

22. *Appeal to Pity:* An attempt to persuade us into buying the car (not by listing its features which would warrant us doing so) but by making us feel sorry for ourselves, and perhaps, a little frightened (by suggesting that we might not be around to do so tomorrow).

28. *Appeal to Fear:* This is still an all-too-common (and, no doubt, effective) campaign appeal. What should sway us, however, are not threats but reasons the speaker may have for why he or she merits our support rather than the opponent.

30. *Personal Attack:* This is a *tu quoque* argument. The fact that Americans also engage in this practice does not make it right, and besides, the protestors are obviously not the ones who do engage in it. If the implication is that the American embassies deserve to be picketed as much as

the Canadian, this is, of course, true—but they no doubt are.

33. *Personal Attack:* This is an example of poisoning the well. Since mediums are the only people who claim to be subject to the phenomena in question, to disparage them in advance in the way done here is to cut ourselves off from the only source of evidence available to us. There is no use, after all, investigating anyone else ("someone more trustworthy"), for they are not subject to the experiences; it is much more reasonable to let the mediums tell us about their experiences and try to see whether they are telling us the truth or not.

35. *Appeal to Ignorance:* The mayor may indeed not have witnessed any dishonesty, but this does not prove there was none. A mayor, no doubt, possesses all sorts of skills but not necessarily those which would uncover athletic infractions.

37. *Personal Attack:* In an abusive *ad hominem* argument, Bishop Wilberforce, in attempting to lower Thomas Huxley in our eyes through ridicule, hoped to lower as well anything Huxley might have said in defense of evolution. Because we are not very inclined to pay serious attention to anyone we do not respect, the attempt may have been effective, though Wilberforce could not be said to have scored a telling hit against the argument for evolution as such.

38. *Appeal to Fear:* This is a not too subtle threat that harm may come to Socrates should he persist in "philosophizing." The threat was not an idle one, for Anytus was indeed one of the three who came to bring charges against Socrates, which resulted in the latter's execution.

49. *Mob Appeal:* This advertisement attempts to sell its product by exploiting such feelings as our love of home and country. It tries to make us feel as if we would be guilty of an act of disloyalty if we did not buy this *American* product.

LOVE IS A FALLACY / Max Shulman

*This fictional account concerns two college students who become
deeply involved with many of the fallacies discussed in part two
of this text, and with one or two others as well. Taken from the
popular novel,* The Many Loves of Dobie Gillis, *the story provides
a delightful, if extreme, illustration of how fallacious reasoning
can affect our everyday lives and a reminder of the limitations of
logical reasoning. In the story, written in the early fifties, readers
may sense overtones of the sexism that was typical of the time.*

Cool was I, and logical. Keen—calculating, perspicacious, acute
and astute—I was all of these. My brain was as powerful as a
dynamo, as precise as a chemist's scales, as penetrating as a scal-
pel. And—think of it!—I was only eighteen.

It is not often that one so young has such a giant intellect. Take,
for example, Petey Bellows, my roommate at the university. Same
age, same background, but dumb as an ox. A nice enough fellow,
you understand, but nothing upstairs. Emotional type. Unstable.
Impressionable. Worst of all, a faddist. Fads, I submit, are the very
negation of reason. To be swept up in every new craze that comes
along, to surrender yourself to idiocy just because everybody else
is doing it—this, to me, is the acme of mindlessness. Not, however,
to Petey.

One afternoon I found Petey lying on his bed with an expression
of such distress on his face that I immediately diagnosed appendi-
citis. "Don't move," I said. "Don't take a laxative. I'll get a doctor."

"Raccoon," he mumbled thickly.

"Raccoon?" I said, pausing in my flight.

"I want a raccoon coat," he wailed.

I perceived that his trouble was not physical, but mental. "Why
do you want a raccoon coat?"

"I should have known it," he cried, pounding his temples. "I
should have known they'd come back when the Charleston came

back. Like a fool I spent all my money for textbooks, and now I can't get a raccoon coat."

"Can you mean," I said incredulously, "that people are actually wearing raccoon coats again?"

"All the Big Men on Campus are wearing them. Where've you been?"

"In the library," I said, naming a place not frequented by Big Men on Campus.

He leaped from the bed and paced the room. "I've got to have a raccoon coat," he said passionately. "I've got to!"

"Petey, why? Look at it rationally. Raccoon coats are unsanitary. They shed. They smell bad. They weigh too much. They're unsightly. They—"

"You don't understand," he interrupted impatiently. "It's the thing to do. Don't you want to be in the swim?"

"No," I said truthfully.

"Well, I do," he declared. "I'd give anything for a raccoon coat. Anything!"

My brain, that precision instrument, slipped into high gear. "Anything?" I asked, looking at him narrowly.

"Anything," he affirmed in ringing tones.

I stroked my chin thoughtfully. It so happened that I knew where to get my hands on a raccoon coat. My father had had one in his undergraduate days; it lay now in a trunk in the attic back home. It also happened that Petey had something I wanted. He didn't *have* it exactly, but at least he had first right on it. I refer to his girl, Polly Espy.

I had long coveted Polly Espy. Let me emphasize that my desire for this young woman was not emotional in nature. She was, to be sure, a girl who excited the emotions, but I was not one to let my heart rule my head. I wanted Polly for a shrewdly calculated, entirely cerebral reason.

I was a freshman in law school. In a few years I would be out in practice. I was well aware of the importance of the right kind of wife in furthering a lawyer's career. The successful lawyers I had observed were, almost without exception, married to beautiful, gracious, intelligent women. With one omission, Polly fitted these specifications perfectly.

Beautiful she was. She was not yet of pin-up proportions, but I felt sure that time would supply the lack. She already had the makings.

Gracious she was. By gracious I mean full of graces. She had an erectness of carriage, an ease of bearing, a poise that clearly indicated the best of breeding. At table her manners were exquisite. I had seen her at the Kozy Kampus Korner eating the specialty of the house—a sandwich that contained scraps of pot roast, gravy, chopped nuts, and a dipper of sauerkraut—without even getting her fingers moist.

Intelligent she was not. In fact, she veered in the opposite direction. But I believed that under my guidance she would smarten up. At any rate, it was worth a try. It is, after all, easier to make a beautiful dumb girl smart than to make an ugly smart girl beautiful.

"Petey," I said, "are you in love with Polly Espy?"

"I think she's a keen kid," he replied, "but I don't know if you'd call it love. Why?"

"Do you," I asked, "have any kind of formal arrangement with her? I mean are you going steady or anything like that?"

"No. We see each other quite a bit, but we both have other dates. Why?"

"Is there," I asked, "any other man for whom she has a particular fondness?"

"Not that I know of. Why?"

I nodded with satisfaction. "In other words, if you were out of the picture, the field would be open. Is that right?"

"I guess so. What are you getting at?"

"Nothing, nothing," I said innocently, and took my suitcase out of the closet.

"Where you going?" asked Petey.

"Home for the week end." I threw a few things into the bag.

"Listen," he said, clutching my arm eagerly, "while you're home, you couldn't get some money from your old man, could you, and lend it to me so I can buy a raccoon coat?"

"I may do better than that," I said with a mysterious wink and closed my bag and left.

"Look," I said to Petey when I got back Monday morning. I threw open the suitcase and revealed the huge, hairy, gamy object that my father had worn in his Stutz Bearcat in 1925.

"Holy Toledo!" said Petey reverently. He plunged his hands into the raccoon coat and then his face. "Holy Toledo!" he repeated fifteen or twenty times.

"Would you like it?" I asked.

"Oh yes!" he cried, clutching the greasy pelt to him. Then a canny look came into his eyes. "What do you want for it?"

"Your girl," I said, mincing no words.

"Polly?" he said in a horrified whisper. "You want Polly?"

"That's right."

He flung the coat from him. "Never," he said stoutly.

I shrugged. "Okay. If you don't want to be in the swim, I guess it's your business."

I sat down in a chair and pretended to read a book, but out of the corner of my eye I kept watching Petey. He was a torn man. First he looked at the coat with the expression of a waif at a bakery window. Then he turned away and set his jaw resolutely. Then he looked back at the coat, with even more longing in his face. Then he turned away, but with not so much resolution this time. Back and forth his head swiveled, desire waxing, resolution waning. Finally he didn't turn away at all; he just stood and stared with mad lust at the coat.

"It isn't as though I was in love with Polly," he said thickly. "Or going steady or anything like that."

"That's right," I murmured.

"What's Polly to me, or me to Polly?"

"Not a thing," said I.

"It's just been a casual kick—just a few laughs, that's all."

"Try on the coat," said I.

He complied. The coat bunched high over his ears and dropped all the way down to his shoe tops. He looked like a mount of dead raccoons. "Fits fine," he said happily.

I rose from my chair. "Is it a deal?" I asked, extending my hand.

He swallowed. "It's a deal," he said and shook my hand.

I had my first date with Polly the following evening. This was in the nature of a survey; I wanted to find out just how much work I had to do to get her mind up to the standard I required. I took her first to dinner. "Gee, that was a delish dinner," she said as we left the restaurant. Then I took her to a movie. "Gee, that was a marvy movie," she said as we left the theater. And then I took her home. "Gee, I had a sensaysh time," she said as she bade me good night.

I went back to my room with a heavy heart. I had gravely under-estimated the size of my task. This girl's lack of information was terrifying. Nor would it be enough merely to supply her with information. First she had to be taught to *think*. This loomed as a project of no small dimensions, and at first I was tempted to give her back to Petey. But then I got to thinking about her abundant physical charms and about the way she entered a room and the way she handled a knife and fork, and I decided to make an effort.

I went about it, as in all things, systematically. I gave her a course in logic. It happened that I, as a law student, was taking a course in logic myself, so I had all the facts at my finger tips. "Polly," I said to her when I picked her up on our next date, "tonight we are going over to the Knoll and talk."

"Oo, terrif," she replied. One thing I will say for this girl: you would go far to find another so agreeable.

We went to the Knoll, the campus trysting place, and we sat down under an old oak, and she looked at me expectantly. "What are we going to talk about?" she asked.

"Logic."

She thought this over for a minute and decided she liked it. "Magnif," she said.

"Logic," I said, clearing my throat, "is the science of thinking. Before we can think correctly, we must first learn to recognize the common fallacies of logic. These we will take up tonight."

"Wow-dow!" she cried, clapping her hands delightedly.

I winced, but went bravely on. "First let us examine the fallacy called Dicto Simpliciter."*

"By all means," she urged, batting her lashes eagerly.

"Dicto Simpliciter means an argument based on an unqualified generalization. For example: Exercise is good. Therefore everybody should exercise."

"I agree," said Polly earnestly. "I mean exercise is wonderful. I mean it builds the body and everything."

"Polly," I said gently, "the argument is a fallacy. *Exercise is good* is an unqualified generalization. For instance, if you have a heart disease, exercise is bad, not good. Many people are ordered by their doctors not to exercise. You must *qualify* the generalization. You must say exercise is *usually* good, or exercise is good *for most people.* Otherwise you have committed a Dicto Simpliciter. Do you see?"

"No," she confessed. "But this is marvy. Do more! Do more!"

"It will be better if you stop tugging at my sleeve," I told her, and

Dicto Simpliciter is Latin for "the fallacy of sweeping generalization."

when she desisted, I continued. "Next we take up a fallacy called Hasty Generalization. Listen carefully: You can't speak French. I can't speak French. Petey Bellows can't speak French. I must therefore conclude that nobody at the University of Minnesota can speak French."

"Really?" said Polly, amazed, *"Nobody?"*

I hid my exasperation. "Polly, it's a fallacy. The generalization is reached too hastily. There are too few instances to support such a conclusion."

"Know any more fallacies?" she asked breathlessly. "This is more fun than dancing even."

I fought off a wave of despair. I was getting nowhere with this girl, absolutely nowhere. Still, I am nothing if not persistent. I continued. "Next comes Post Hoc. Listen to this: Let's not take Bill on our picnic. Every time we take him out with us, it rains."

"I know somebody just like that," she exclaimed. "A girl back home—Eula Becker, her name is. It never fails. Every single time we take her on a picnic—"

"Polly," I said sharply, "it's a fallacy. Eula Becker doesn't cause the rain. She has no connection with the rain. You are guilty of Post Hoc if you blame Eula Becker."

"I'll never do it again," she promised contritely. "Are you mad at me?"

I sighed. "No, Polly, I'm not mad."

"Then tell me some more fallacies."

"All right. Let's try Contradictory Premises."

"Yes, let's," she chirped, blinking her eyes happily.

I frowned, but plunged ahead. "Here's an example of Contradictory Premises: If God can do anything, can He make a stone so heavy that He won't be able to lift it?"

"Of course," she replied promptly.

"But if He can do anything, He can lift the stone," I pointed out.

"Yeah," she said thoughtfully. "Well, then I guess He can't make the stone."

"But He can do anything," I reminded her.

She scratched her pretty, empty head. "I'm all confused," she admitted.

"Of course you are. Because when the premises of an argument contradict each other, there can be no argument. If there is an irresistible force, there can be no immovable object. If there is an immovable object, there can be no irresistible force. Get it?"

"Tell me some more of this keen stuff," she said eagerly.

I consulted my watch. "I think we'd better call it a night. I'll take you home now, and you go over all the things you've learned. We'll have another session tomorrow night."

I deposited her at the girls' dormitory, where she assured me that she had had a perfectly terrif evening, and I went glumly home to my room. Petey lay snoring in his bed, the raccoon coat huddled like a great hairy beast at his feet. For a moment I considered waking him and telling him that he could have his girl back. It seemed clear that my project was doomed to failure. The girl simply had a logic-proof head.

But then I reconsidered. I had wasted one evening; I might as well waste another. Who knew? Maybe somewhere in the extinct crater of her mind a few embers still smoldered. Maybe somehow I could fan them into flame. Admittedly it was not a prospect fraught with hope, but I decided to give it one more try.

Seated under the oak the next evening I said, "Our first fallacy tonight is called Ad Misericordiam."

She quivered with delight.

"Listen closely," I said. "A man applies for a job. When the boss asks him what his qualifications are, he replies that he has a wife and six children at home, the wife is a helpless cripple, the children have nothing to eat, no clothes to wear, no shoes on their feet, there are no beds in the house, no coal in the cellar, and winter is coming."

A tear rolled down each of Polly's pink cheeks. "Oh, this is awful, awful," she sobbed.

"Yes, it's awful," I agreed, "but it's no argument. The man never answered the boss's question about his qualifications. Instead he appealed to the boss's sympathy. He committed the fallacy of Ad Misericordiam. Do you understand?"

"Have you got a handkerchief?" she blubbered.

I handed her a handkerchief and tried to keep from screaming while she wiped her eyes. "Next," I said in a carefully controlled tone, "we will discuss False Analogy. Here is an example: Students should be allowed to look at their textbooks during examinations. After all, surgeons have X rays to guide them during an operation, lawyers have briefs to guide them during a trial, carpenters have blueprints to guide them when they are building a house. Why, then, shouldn't students be allowed to look at their textbooks during an examination?"

"There now," she said enthusiastically, "is the most marvy idea I've heard in years."

"Polly," I said testily, "the argument is all wrong. Doctors, lawyers, and carpenters aren't taking a test to see how much they have learned, but students are. The situations are altogether different, and you can't make an analogy between them."

"I still think it's a good idea," said Polly.

"Nuts," I muttered. Doggedly I pressed on. "Next we'll try Hypothesis Contrary to Fact."

"Sounds yummy," was Polly's reaction.

"Listen: If Madame Curie had not happened to leave a photographic plate in a drawer with a chunk of pitchblende, the world today would not know about radium."

"True, true," said Polly, nodding her head. "Did you see the movie? Oh, it just knocked me out. That Walter Pidgeon is so dreamy. I mean he fractures me."

"If you can forget Mr. Pidgeon for a moment," I said coldly, "I would like to point out that the statement is a fallacy. Maybe Madame Curie would have discovered radium at some later date. Maybe somebody else would have discovered it. Maybe any number of things would have happened. You can't start with a hypothesis that is not true and then draw any supportable conclusions from it."

"They ought to put Walter Pidgeon in more pictures," said Polly. "I hardly ever see him any more."

One more chance, I decided. But just one more. There is a limit to what flesh and blood can bear. "The next fallacy is called Poisoning the Well."

"How cute!" she gurgled.

"Two men are having a debate. The first one gets up and says, 'My opponent is a notorious liar. You can't believe a word that he is going to say.' . . . Now, Polly, think. Think hard. What's wrong?"

I watched her closely, as she knit her creamy brow in concentration. Suddenly a glimmer of intelligence—the first I had seen— came into her eyes. "It's not fair," she said with indignation. "It's not a bit fair. What chance has the second man got if the first man calls him a liar before he even begins talking?"

"Right!" I cried exultantly. "One hundred per cent right. It's not fair. The first man has *poisoned the well* before anybody could drink from it. He has hamstrung his opponent before he could even start. . . . Polly, I'm proud of you."

"Pshaw," she murmured, blushing with pleasure.

"You see, my dear, these things are not so hard. All you have to do is concentrate. Think—examine—evaluate. Come now, let's review everything we have learned."

"Fire away," she said with an airy wave of her hand.

Heartened by the knowledge that Polly was not altogether a cretin, I began a long, patient review of all I had told her. Over and over and over again I cited instances, pointed out flaws, kept hammering away without letup. It was like digging a tunnel. At first everything was work, sweat, and darkness. I had no idea when I would reach the light, or even *if* I would. But I persisted. I pounded and clawed and scraped, and finally I was rewarded. I saw a chink of light. And then the chink got bigger and the sun came pouring in and all was bright.

Five grueling nights this took, but it was worth it. I had made a logician out of Polly; I had taught her to think. My job was done. She was worthy of me at last. She was a fit wife for me, a proper hostess for my many mansions, a suitable mother for my well-heeled children.

It must not be thought that I was without love for this girl. Quite the contrary. Just as Pygmalion loved the perfect woman he had fashioned, so I loved mine. I decided to acquaint her with my feelings at our very next meeting. The time had come to change our relationship from academic to romantic.

"Polly," I said when next we sat beneath our oak, "tonight we will not discuss fallacies."

"Aw, gee," she said, disappointed.

"My dear," I said, favoring her with a smile, "we have now spent five evenings together. We have gotten along splendidly. It is clear that we are well matched."

"Hasty Generalization," she repeated. "How can you say that we are well matched on the basis of only five dates?"

I chuckled with amusement. The dear child had learned her lessons well. "My dear," I said, patting her hand in a tolerant manner, "five dates is plenty. After all, you don't have to eat a whole cake to know that it's good."

"False Analogy," said Polly promptly. "I'm not a cake. I'm a girl."

I chuckled with somewhat less amusement. The dear child had learned her lessons perhaps too well. I decided to change tactics. Obviously the best approach was a simple, strong, direct declaration of love. I paused for a moment while my massive brain chose the proper words. Then I began:

"Polly, I love you. You are the whole world to me, and the moon and the stars and the constellations of outer space. Please, my darling, say that you will go steady with me, for if you will not, life will be meaningless. I will languish. I will refuse my meals. I will wander the face of the earth, a shambling, hollow-eyed hulk."

There, I thought, folding my arms, that ought to do it.

"Ad Misericordiam," said Polly.

I ground my teeth. I was not Pygmalion; I was Frankenstein, and my monster had me by the throat. Frantically I fought back the tide of panic surging through me. At all costs I had to keep cool.

"Well, Polly," I said, forcing a smile, "you certainly have learned your fallacies."

"You're darn right," she said with a vigorous nod.

"And who taught them to you, Polly?"

"You did."

"That's right. So you do owe me something, don't you, my dear? If I hadn't come along you never would have learned about fallacies."

"Hypothesis Contrary to Fact," she said instantly.

I dashed perspiration from my brow. "Polly," I croaked, "you mustn't take all these things so literally. I mean this is just classroom stuff. You know that the things you learn in school don't have anything to do with life."

"Dicto Simpliciter," she said, wagging her finger at me playfully.

That did it. I leaped to my feet, bellowing like a bull. "Will you or will you not go steady with me?"

"I will not," she replied.

"Why not?" I demanded.

"Because this afternoon I promised Petey Bellows that I would go steady with him."

I reeled back, overcome with the infamy of it. After he promised, after he made a deal, after he shook my hand! "The rat!" I shrieked, kicking up great chunks of turf. "You can't go with him, Polly. He's a liar. He's a cheat. He's a rat."

"Poisoning the Well," said Polly, "and stop shouting. I think shouting must be a fallacy too."

With an immense effort of will, I modulated my voice. "All

right," I said. "You're a logician. Let's look at this thing logically. How could you choose Petey Bellows over me? Look at me—a brilliant student, a tremendous intellectual, a man with an assured future. Look at Petey—a knothead, a jitterbug, a guy who'll never know where his next meal is coming from. Can you give me one logical reason why you should go steady with Petey Bellows?"

"I certainly can," declared Polly. "He's got a raccoon coat."

SUGGESTED READINGS FOR PART TWO

The works listed here should prove interesting to readers following the study of part two, although some are more advanced than the books suggested at the conclusion of part one. With the exception of the first, they are available in inexpensive paperback editions.

C. L. Hamblin. *Fallacies.* London: Methuen, 1972. A chronological account, beginning with Aristotle, of the history of common fallacies. The author is a professor of philosophy at the University of New South Wales.

Wilson Bryan Key. *Media Sexploitation.* Englewood Cliffs, N.J.: Prentice-Hall, 1976. An account of how the mass media arouses and manipulates us by playing on our fantasies, fears, and intimate habits into buying whatever it decides to offer us. Key explored this theme in an earlier book entitled *Subliminal Seduction* (1973) and more recently still in *Clam-Plate Orgy* (1980).

Stanley Milgram. *Obedience to Authority.* New York: Harper & Row, 1974. Based on research carried out at Yale University, this account of a famous experiment explores what it is about the nature of obedience to authority that can create a situation in which one person can command another to harm, even destroy, an innocent third, and have those commands carried out impassively, submissively, routinely.

David Ogilvy. *Confessions of an Advertising Man.* New York: Dell, 1963. This is one of the first works devoted to exposing the all too successful methods exploited by advertisers to sell their products.

Lionel Ruby. *The Art of Making Sense.* 3rd ed. Philadelphia: J. B. Lippincott, 1974. A delightful account of the art of thinking by the late and beloved American teacher of philosophy.

Gilbert Ryle. *Dilemmas.* Cambridge, Eng.: At the University Press, 1960. How certain perennial problems can be resolved by application of the method of linguistic analysis.

Michael Wheeler. *Lies, Damn Lies, and Statistics: The Manipulation of Public Opinion in America.* New York: Dell, 1976. As this book shows, not only our buying preferences but our political and social opinions can be—and often are—"managed."

Ludwig Wittgenstein. *The Blue and Brown Books.* 2nd ed. New York: Harper & Row, 1969. An account of how various philosophical perplexities have their source in linguistic confusions. Written by the founder of one of the most important contemporary movements of philosophy, namely, analytic philosophy.

"So, then . . . Would that be 'us the people' or 'we the people'?"

Appendix

Writing with Clarity and Reason

The task of writing an essay allows you to practice and test the skills you learn in your study of logical reasoning; in addition, we might say that writing an essay lets you argue with other writers. The essay is, at heart, an extended argument; in it, a writer makes an assertion about a subject (the assertion is often referred to as the *conclusion*) by reasoning through specific examples (the *premises* that support a conclusion). Your task as a student is to practice and test the reasoning of other writers. When you understand how the writers you read come to a particular conclusion, you can agree with, contest, or modify their arguments with a measure of authority and confidence. If you write your response as an essay, you can hold it out as a gesture of credibility, a sign that you too can reason through premises to a conclusion. You won't simply be summarizing the reading you do; you will be demonstrating that you have thought about what you have been studying and reading.

To take charge of your ideas in this way is a challenge; however, writing a coherent, thoughtful essay is a craft. Like other crafts, it can, with practice, be mastered. As you work on developing your

writing skills, you should be reading *With Good Reason* and studying the ways in which specific fallacies apply to arguments you
both read and write about. However, you should also keep in mind
that writing an essay involves more than an understanding of
what is either good or fallacious reasoning. The creation of essay
demands an analysis of your subject, a consideration of the language you will use to talk about your subject, and an assessment
of the audience for whom you write.

To be more more precise, an essay comprises logic, grammar,
and rhetoric. *Logic* refers to the reasoning process you will be
exercising throughout this book; specifically, to an examination of
the relationship between sentences that knits them together as a
unit of thought. As you will learn from reading *With Good Reason,*
logic involves premises *(All men are mortal. Socrates was a man.)*
and a conclusion derived from those premises *(Socrates was mortal.). Grammar* refers to the structure of sentences. For instance,
the brief sentence *Socrates was mortal* is broken down grammatically as follows:

> Socrates   (subject of sentence)
>   was   (past tense of verb *be*)
>   mortal   (adjective modifying/describing
>   subject)

Rhetoric refers to the ways in which we manipulate logic and
grammar to provoke, interest, or persuade a reader. For instance,
the conclusion *Socrates was mortal* seems legitimate, and its
grammar is straightforward. However, an essayist might play
with the logic and syntax of the Socrates syllogism to arrive at a
related but less easily provable conclusion:

> We see that all men are mortal, that even Socrates was no
> exception. Socrates is lost to us; we know only Plato's inter-
> pretations of Socrates' words. Nothing of what we are lives
> forever. We die, and though our thoughts might live after us
> in books, though our biographies might be written and re-
> written, we cannot speak for ourselves. Our mortality means
> that other men and women must speak for us; in losing our
> bodies we lose our voices.

Note how the above passage makes use of logic and grammar in order to reach a new conclusion, one that relies as much on eloquence and style as on grammatically correct sentences and sound reasoning. In this case, short, terse statements alternate with longer, more elaborate ones, as the argument links Socrates and all men to arrive at an emotionally provocative assertion about mortality. The writer of this passage is concerned with persuading an audience; with that goal in mind, the writer expands a logically coherent conclusion to include a more ambiguous and emotional assertion, and varies the syntax to create a persuasive voice. The result establishes a mood and tone that can sway an audience. From this example, you can see that logic, sentence structure, and techniques of persuasion interweave in the course of writing an argumentative essay.

1. AN OVERVIEW OF ESSAY STRUCTURE

Let's assume that you have been assigned to write a three- to five-page essay for your class. Such an assignment can be guided by a fairly straightforward structure. Although essays assume various shapes and sizes, the pattern that will be easiest for you to follow (and to modify if you so choose) will be the traditional three-part essay. The following description of this structure, along with the student essay used to demonstrate it, conforms also to certain conventions of *argumentative* writing. An argumentative essay takes a widely held conclusion about a subject, or a conclusion particular to another writer or thinker, and agrees with, disagrees with, or somehow modifies that conclusion by presenting premises and supporting statements organized into logically coherent paragraphs. The structure of such an essay is as follows.

BEGINNING

The beginning of the traditional essay serves to introduce the subject you intend to write about. In addition, it states the *conclusion* (also known as the *thesis* of an essay) you intend to prove in the course of your essay. Thus, the beginning paragraph of an essay is often referred to as the *introductory* or *thesis* paragraph. Let's see how one student writer begins an essay on Socrates:

> In 399 B.C., a man named Socrates was brought to trial and condemned to death by the very society that he said he had been sent to by God to help and improve. The charges brought against him were twofold. He was accused, first, of impiety, or "not believing in the gods in whom the city believes, but in other new divinities" (*Apology* 24B). Second, he was accused of corrupting the minds of the young. Socrates was tried on these counts by a jury of his peers, and his conviction reflects that he did transgress certain laws upholding Athenian society. It is possible, however, to discover a more complicated set of reasons for his death, namely, that the Athenians felt personally and culturally threatened by Socrates' teachings.

This writer's beginning paragraph moves through a process that (1) establishes that her subject is Socrates' trial and condemnation, (2) narrows and focuses that subject to the charges raised against him at his trial, and (3) lets her readers know she intends to disagree with the conclusion that Socrates was condemned to death merely because he broke the law; she advises that her focus will be on the Athenian response to the content of Socrates' teachings. Such a beginning both prepares the reader for the subject being discussed *and* narrows the subject so that it is clearly defined. Her *conclusion,* or thesis statement, is clearly articulated; as readers, we are ready for a discussion of why she questions and considers important the "facts" of the situation surrounding Socrates' death.

THE MIDDLE, OR BODY

Each paragraph in an essay's middle section presents and supports the premises with details and examples (often called *evidence*) supporting the essay's conclusion. You can get a general sense of the construction of the body of an essay by looking over the next four paragraphs of the student essay, which the writer devotes to developing and supporting her questioning of Socrates' death:

In order to understand the Athenians' decision, we must first understand the circumstances surrounding the trial, as well as be familiar with Socrates' character and convictions. Plato's *Apology,* our most reliable source for understanding Socratic philosophy, offers a demonstration of Socrates' values and his methods of reasoning. Socrates defends himself with logic and dignity, addressing and refuting every accusation made against him, as well as some that had been left unspoken. As he states his case, Socrates again and again assures the jury that what he is saying is, indeed, the truth, all the while inferring, not always subtly, that his accusers are lying. In addition, he continually refers to his activities as a duty or religious obligation assigned to him by God. According to Plato's interpretation of the trial, Socrates feels that what God has ordained, no person can condemn. Not once, even after he has been sentenced to death, does Socrates appeal to the jurors' pity or attempt to flatter them; he even goes so far as to say that, if he were acquitted, he would continue the activities that had brought him to trial. Socrates, then, was a man with a religious mission who would not be budged by the Athenians' opinion that he was breaking the laws that governed the city.

This second paragraph sets up a context for understanding the writer's argument, in that it changes the image of Socrates as a man "accused of impiety" into one of Socrates as "a man of conviction." Although the writer sets forth the subject and conclusion in the essay's beginning, in the body of the essay she continues to define the subject by presenting a particular view of Socrates' motivations and behavior. In so doing, the writer narrows the subject so that specific examples can be offered to shape a convincing argument. The rest of the middle section is firmly based in the explicitly stated assumption that Socrates obeyed the laws of God rather than the laws of society:

Just what were the activities that made Socrates so unpopular with so many? Socrates believed that he was a gift from God to the people of Athens, sent to open their eyes to the faults in both themselves and their city, in order that they might correct them. Socrates' method of doing so was this: upon meeting a man who saw himself as wise, Socrates would, as Plato states, "come to the assistance of the god and show him that he [the man] is not wise" (23C). He would thus question the man about his beliefs and assertions, forcing him to see that his opinions were full of contradictions and inconsistencies. Socrates felt that if people acknowledged their ignorance, they would then be open to searching out true knowledge; in Socratic thought, people who deny their ignorance cannot act and think as God wishes them to. However, Socrates' continual challenge to the Athenian community caused its citizens to condemn both the philosopher and his philosophy.

It seems, however, that the difficulty lay not so much with Socrates as with the Athenians themselves. On an individual level, an Athenian probably felt embarrassed and humiliated when Socrates posed questions that revealed flaws and inconsistencies in his or her ability to think well. His questions made the people he spoke with uncomfortably aware of their ignorance, as well as appear foolish, oftentimes in front of other people. Take, for instance, his response to the judges regarding the charge that he was impious, or arrogant toward God: "What is probable, gentlemen, is that in fact that god is wise . . . and that when he says this man, Socrates, he is using my name as an example, as if he said: 'This man among you, mortals, is wisest who, like Socrates, understands that his wisdom is worthless'" (23B). In this way, Socrates could calmly destroy the basis of the Athenians' assumption that Socrates thought himself above God; at the same time he showed himself to be an extremely pious person. Such an attitude could only have made the judges feel foolish, so that they, like Euthyphro in Plato's dialogue, longed to escape further humiliation.

On a larger scale, Socrates challenged the structure of Athenian society, pointing out flaws and weaknesses that its citizens had closed their eyes against. Athenians soon wished to rid themselves of this aggravating man who persisted in exposing their personal failings and disrupting an established social code. To make matters worse, Socrates pointed out problems without suggesting solutions. If he had offered an alternative, the people might have found his teachings easier to swallow, or at least found concrete reasons to bring him to trial. Instead, Socrates persisted only in awakening Athenians to their faults so that they, in turn, might rectify them. Socrates likened the Athenians to a drowsing horse that can be roused and motivated to action by a tiny stinging fly (30E). The Athenians became annoyed with this stinging fly, Socrates, and swatted it dead.

At this point, the first part of this essay's middle section ends. The four paragraphs raise and support two premises: first, that Socrates was a man deeply committed to following God's laws, and second, that the Athenians found this philosophy and Socrates' manner of presenting it personally and (by extension) culturally threatening. Thus, this writer arrives at a preliminary proof of her conclusion, in that she presents premises and supporting statements refuting the idea that Socrates was simply a criminal.

However, this essay's middle section must continue if it is to justify yet another premise underpinning its argument: that the Athenians "shut their eyes" to the flaws in both their reasoning processes, and by extension, to flaws in Athenian culture. Such an assumption cannot be proven beyond doubt; with that in mind, this writer presents a transitional paragraph that appeals to our acceptance of a "truth" about human nature, and then offers two analogies in support of that truth:

Is this reluctance to seek improvement in self and society characteristic only of the Athenians, or does it have wider connotations? Would, perhaps, a different society respond differently under similiar circumstances? To answer this question we must determine if the traits exhibited by Socrates' contemporaries are peculiar to their society, or if they are timeless characteristics.

We can approach this problem by first asserting that like the Athenians, most people would rather not see their own shortcomings and weaknesses. Take, for example, the case of the alcoholic. It is common to say that the biggest step in the recovery of an alcoholic is getting the sick individual to recognize his or her problem and to seek help. Such a step is considered "the biggest" for a very good reason: many people with drinking problems are unable to admit or to even realize that they cannot control their drinking. They argue that they could stop drinking at any time. They do not want to face the problem, even when they see their alcoholism destroying careers, marriages, and families. In short, they do not see that they live a life of self-deception. Though they could learn to recognize their condition, they fail to do so.

It might be said, then, that many people do not want their basic convictions, including the society with which they identify, to be disputed. As a house is built upon a foundation, so people build their lives on these convictions, and if the house's foundation is shaken, the house itself might topple. Although people welcome some kinds of change, they are afraid of the sort of change that calls for a complete readjustment of their lives and thinking patterns; the mere suggestion of such a thing seems threatening, even personally demeaning.

Thus, any man or woman who plants seeds of doubt in people's minds as to the rightness of their suppositions, and especially one who offers people no alternative course of action, is bound to make enemies, no matter what society he or she may belong to. It seems likely that Socrates would have been rejected by his social community no matter when or where he was born (though his sentence might have varied, depending on the culture that tried him). People would resent his attempts to help them improve themselves whether they lived in Athens, 399 B.C., or New York City, 1989.

You have seen that the middle section of this essay is composed of two parts: the first part reinterprets Socrates' beliefs and actions and the Athenian response to those actions, while the second part offers a plausible, though disputable, explanation of human motivation and behavior. The middle section, then, is the heartbeat of this argument, as it is with any argument. Its structure can vary, but it always should provide careful, point-by-point premises and support for the conclusion presented in an essay's beginning. In addition, the middle section should establish the writer's specific concerns, the way that this writer's middle section bases the explanation in a clearly stated interpretation of the motivations of Socrates and the Athenians for acting as they did.

ENDING

The end of an essay functions as a summation of the conclusion and premises presented in the beginning and middle sections; the summation should pull together the pieces of the puzzle the essay offers a reader. For instance, our student essay's ending draws together the two parts of her middle section so that they function as a final support of her original conclusion:

> Why, then, was Socrates condemned? The stated accusations against him do not appear to explain why he lost his life. These accusations were merely formalities that let his opponents build a legal argument against his activities and thus bring him to court. Socrates did not die because he was impious or because he corrupted the minds of the young. Rather, he died because of the unspoken charges against him. These charges, like the given ones, were twofold. First, Socrates exposed the weakness and ignorance in individual Athenian citizens, thereby incurring their anger. Second, his rigorous questioning of people eventually threatened the security of Athenian society as a whole, thereby causing the Athenian community to attack him. He was destroyed by the very society he sought to improve, because the members of that society refused to see that they might need improvement. The Athenians were not ready for one such as Socrates, and it seems doubtful that any person or community who treats criticism as a threat ever will be ready.

This ending efficiently and *thoughtfully* summarizes the writer's conclusion and the premises supporting that conclusion. This is to say that an essay's ending works as more than a restatement of its major points; the ending also makes a final effort to convince or persuade the reader of the conclusion, and relies on the premises presented in the middle section to help the reader come to a final understanding of the writer's argument.*Now that you've examined one essay with an eye to its structure, you should have some idea of the kind of argument you will be presenting to your reader. It is never premature to think about structure, since the type of essay you write will shape your presentation of ideas. If you approach your writing task with knowledge of the writing conventions you'll be working with, you will find that your argument will develop more easily than if you work with no sense of the form you want to use.

*Plato. *The Trial & Death of Socrates.* Trans. G. M. A. Grube. Indianapolis: Hackett Publishing Company, 1975.

2. DEVELOPING YOUR ESSAY

Both student and professional writers sometimes succumb to procrastination, for the effort required to write a grammatically correct, well-reasoned, and persuasive essay can seem daunting. However, if you ask yourself some questions and follow a few simple rules, you will find that you can plan your writing task so that it becomes a manageable one.

FINDING YOUR TOPIC

It is likely that your essay assignment will offer you some focus for determining the topic about which you will write, in that your instructor will probably ask you to look at a specific book or think about a particular idea you've covered in class. As you begin to narrow possible topics, bring to mind a fairly standard definition of argument: language that persuades. This means that your essay will try to convince your reader that a particular viewpoint about a particular subject is plausible. Whether you write about reductions in nuclear weapons or shoe styles for tall men, you will need to define your subject so that you arrive at an *arguable topic*. Chapter 1 offers a detailed discussion of the components of sound arguments, but the following suggestions may help you define one for your essay.

An arguable topic should try to persuade readers of something, change their minds about something, or get them to do something. For instance, the student essay we looked at earlier attempts to change our minds about the reasons why Socrates was condemned to death, and to persuade us of an alternative explanation for his rejection by the Athenian community. The writer does not simply "report" information about Socratic philosophy; she actively uses that information to construct her interpretation of the reasons for his conviction.

An arguable topic should discuss a problem without an easy solution, or ask a question that has no absolute answer; one of the reasons you are writing an essay is to demonstrate your ability to *think* about a subject, not merely to summarize information you've read. The student writer chose a topic that could not be justly addressed by a dip into an encyclopedia; instead, she discussed Socrates' death in terms of a consideration of Plato's *Apology* and some examples of and assumptions about the way humans respond to criticism.

An arguable topic should not be based solely on personal conviction. To a certain extent, any argument represents a writer's opinion on a subject; indeed, argumentative essays would not exist if writers did not bring personal experience and observation to their writing tasks. In addition, essays are usually inductive in their approach to a problem, which means that they rely more on experience and observation than do deductive approaches. However, your written argument must conform to certain conventions of good reasoning if your argument is to be sound. For instance, the student writer has a definite opinion about the subject, and the ending makes an authoritative statement about the conclusion that's been asserted: "The Athenians were not ready for one such as Socrates, and it seems doubtful that any person or community who treats criticism as a threat ever will be ready." However, the writer takes care to base this opinion in an explanation that is illustrated by specific, easily understandable examples. The writer also uses Plato's text as an authoritative reference for the argument, thereby looking to primary sources to support the assertion that human weakness, more than the law, led to Socrates' conviction. Finally, the example of the alcoholic presents the reader with a specific illustration of the assumption that "most people would rather not see their shortcomings and weaknesses." You can see, then, that writers may begin with a "hunch" that their ideas are sound, but must reach outside themselves in order to persuade a reader.

An arguable topic should offer a position that readers could realistically disagree with. This last point is crucial to the development of an essay. Your task as a writer is not to present "Truth" with a capital T, but rather to offer a well-supported explanation of your viewpoint on a subject. That is why, on student essays, instructors will often mark or circle statements such as *there is no doubt, it is absolutely certain,* or *the right way to think about this situation.* Such phrases belong, if anyplace, only in the courtroom, where rhetoric is used to persuade a jury that only one verdict is the correct one. Within the context of an essay, you cannot offer a final answer to a controversy, and if you try to discuss a subject as though you own its only correct conclusion, you will very probably commit the fallacies discussed in this book, and end up with an essay that will not convince a discerning reader. Writing an argumentative essay does not mean you fight an idea like a gladiator, but rather that you offer the strongest support possible for one of many perspectives on a subject.

Keeping in mind the criteria discussed above, let's look at a sentence that represents an inarguable topic, and revise it so that it becomes legitimately arguable.

> The government should repeal laws that prohibit citizens from keeping guns in their houses. My uncle was killed because he could not defend himself against a burglar.

The statement is based too much in personal opinion, and insists on a cause/effect relationship that might not be true; although it is possible that the writer's uncle was killed because he did not own a gun, the consequence does not necessarily follow from the presented facts. Such an illustration offers a potentially persuasive example to an argument, but needs to be prefaced by a more focused and objective statement. Look now at this revision:

> Many good reasons exist for the repeal of gun control laws. For the purposes of my argument, the primary reason is that responsible people should be permitted to protect themselves. My own uncle was killed in his bedroom by a burglar with a gun. It is possible that he might have been able to defend himself had he possessed a weapon. Such a possibility, along with several others, fuels my sense that gun control harms rather than helps the private citizen.

Although the writer's primary example is still based in personal experience, it now functions as an illustration of the more general assumption that private citizens should be permitted to protect themselves. In addition, the writer takes care to explain that he is presenting a point of view that is open to questioning. Such an explicitly defined position invites the reader to explore the argument, rather than demands that the reader agree with it.

SPECIFYING YOUR TOPIC

You may already use techniques such as brainstorming, freewriting, and list-making when you begin generating ideas for an essay. Such activities are known as *prewriting,* meaning that they are part of the work you do before you actually begin writing a first draft of your essay. If you currently do not use any system to develop your topic, look over the following brief descriptions of prewriting strategies. In addition to helping you find a focus for your topic, these prewriting activities will give you information about the premises and examples you will use to support your argument.

Brainstorming
You can begin brainstorming about a topic by discussing your assignment with your instructor and with friends; by, for example, proposing ideas, asking questions, and observing what aspects of the subject spark discussion of (and sometimes challenges to) your point of view. Brainstorming with a pen involves sitting down and writing for a specified amount of time, say twenty minutes, about the subject you're considering. You should write down everything you can think of about your subject; use questions, a list, or a rough outline. You can even use sentence fragments that make sense to no one but you. When you finish, look over what you've written to find recurring ideas, pieces of evidence that might support premises, perhaps even a central point that could serve as a working conclusion.

Freewriting
Freewriting is very similar to brainstorming, except that it requires you to write without stopping and in complete sentences. Again, you should write down everything you can think about your potential topic, but in addition, you should write down *anything* you think about so as not to stop writing. Often, seemingly random associations will help you uncover premises that help support your argument. (Remember that an argumentative essay often begins as an intuitive hunch about a subject.)

Questioning
An argumentative essay asks that you persuade or convince a reader of your viewpoint. The following questions may help you

analyze and focus your topic so that your conclusion and premises seem plausible to a reader:

1. What conclusion are you making about your topic?
2. What specific examples support your conclusion?
3. What underlying assumptions support the examples for your conclusion?
4. What backup evidence can you find to add more support to your conclusion?
5. What disagreements might be raised about your conclusion?
6. In what ways should you qualify your conclusion?

CONSTRUCTING A WORKING CONCLUSION

Many writers work with the strategies listed above and end up with statements that are arguable, but too vague to be encompassed within a short essay. A useful technique for focusing your subject involves taking a very general statement and revising it again and again until it becomes a very specific one:

Socrates was condemned because his society refused to acknowledge that he was right.

1. Socrates was condemned *to death* because *the Athenians felt threatened by his teachings.*
2. Socrates was condemned to death *not simply because he broke the law, but* because the Athenians *found his religious convictions, along with his questioning of their wisdom, a personal and cultural threat.*
3. Socrates was condemned to death not simply because he broke the law, but also because the *Athenian community felt threatened by his questioning of their wisdom and values.*

You may end up with very long sentences that seem to ramble endlessly about your topic. Don't worry about writing correct sentences at this point; instead, work to clarify vague or general phrases so that you learn the precise limits of your topic.

EXERCISE

Revise the following statements until they are as specific as you can make them. Do not limit your thoughts to the information contained in each statement; use any information you have about the subject described to tighten and clarify its focus.

1. Homeless people present America with a big problem.
2. Drug abuse presents America with a big problem.
3. Family life should offer people security and affection.
4. Gun-control laws should be lifted.
5. Presidential elections are run by television broadcasters.
6. Abortion is every woman's right.
7. Writing is hard work.

RHETORICAL CONSIDERATIONS

As you brainstorm and create working conclusions, you should be considering the voice or tone you want to assume in your essay. Too often, writers get trapped within the rules governing a traditional essay. They dutifully create conclusions, write essays with beginnings, middles, and endings, but don't permit themselves to take any pleasure in the fact that they are writing as authorities on a subject, and working to reach an audience. Many students insist that the audience is really the instructor and the red marking pen; however, while it is true that your work will be read and evaluated by your instructor, it's also the case that your instructor would probably like to read an essay that you enjoyed writing. A major purpose of your writing task is to present your conclusion, premises, and supporting statements so that they make for interesting and engaging writing and reading experiences. The rhetoric of the essay demands an imaginative consideration of who you are as a writer, and who you are trying to persuade with your writing. John R. Trimble offers some useful points to use in thinking about audience:

> What [writing] involves is one person earnestly attempting to communicate with another. Implicitly, then, it involves the reader every bit as much as the writer, since the success of the communication depends solely on how the reader re-

ceives it. Also, since more than one person is involved, and since all people have feelings, it has to be as subject to the basic rules of good manners as any human relationship. . . . Far from writing in a vacuum, [the writer] is conversing, in a very real sense, with another human being, just as I am conversing right now with you, even though that person—like you—may be hours, or days, or even years away from him in time. (*Writing with Style: Conversations on the Art of Writing.* Englewood Cliffs, N.J.: Prentice-Hall, 1975, p. 15)

Trimble goes on to state that the requirements of talking with a reader include writing with clarity and reason, on a subject that is legitimately arguable and tightly focused. He insists, however, that you also consider the following aspects of rhetoric when writing an essay.

1. Phrase your thoughts clearly so that [they are] easy to follow.
2. Speak to the point so that you don't waste your reader's time.
3. Anticipate [your reader's] many questions and responses.
4. Offer . . . variety and humour to keep [your reader] interested.
5. Converse with [your reader] in a warm, friendly, open manner instead of pontificating like a self-important pedant. (p. 19)

In sum, Trimble is telling writers to say what they mean so that they can be understood. However, though the idea of writing in a courteous and clear manner seems straightforward, it can become terribly confusing when it is mixed with the conventions governing the traditional scholarly essay. Student writers are told to write in formal rather than colloquial English, to use many logical connectives (*however, thus, in addition, on the one hand, since, because, in conclusion*), and never to use the first-person singular. These rules do represent good advice, but only if they are used to help a reader make sense of the logic and rhetoric of an essay. Otherwise, the writer can feel less in control of what he or she is saying, and so lose control of the essay's logic. For instance, although the student essay we examined earlier was written to satisfy a course requirement, its author does not simply fling about big words to impress her reader. Instead, she chooses her language to create a sophisticated voice that both explains and persuades:

> In order to understand the Athenians' decision, we must first understand the circumstances surrounding the trial, as well as be familiar with Socrates' character and convictions. Plato's *Apology,...* offers a demonstration of Socrates' values and his methods of reasoning.

This statement establishes the writer as an authority on the subject of Socrates, but note how her use of *we* implicitly asks the reader to join her in an exploration of her interpretation. In addition, her reference to Plato establishes her legitimate attempt to join the scholars who think and write about Socrates. Finally, she does not use "filler" phrases; every word is relevant to the point her essay is trying to make. Even her use of the phrase *in order to understand* (like *we,* this phrase is a formality, a polite gesture) establishes a transition between the beginning and middle of her essay and allows her reader to participate in the reasoning process.

AUDIENCE AND WORD CHOICE

In his essay "The Well of English, Now Defiled" (*Princeton Alumni Weekly,* September 26, 1958), Willard Thorp coins the term "No-English" to describe the problem of writers who use language that obscures the ideas they try to express. The problem seems to be based in the incorrect choice of individual words, in the use of trite or hackneyed expressions, and in the writing of overly long, verbose sentences. "No-English" represents a problem in writing that you can easily correct; you need only spend some time considering what, precisely, you are trying to say, and with what audience you are trying to communicate. If you think hard about your idea and about your audience, you can then adapt the three-part structure and use it to talk with, rather than to talk at, your reader.

Let's look at two sentences that demonstrate the problem of "No-English," and revise them so that they express the main thought simply and coherently.

Socrates a man of great concern held a strong conviction by allowing his duty to the god. Some people did see that Socrates was a man of good knowledge of skill and was qualified by teaching and applying his knowledge to his people.

In terms of rhetoric, the writer needs to simplify the use of language so that the reader understands what is being said (for instance, what is "a man of great concern"?). In terms of logic, the two sentences do not meaningfully relate to each other, and need to be revised so that they work together to express a thought. In terms of grammar, the writer needs to use correct punctuation if the sentences are to conform to conventions of standard English. Let's look at the revision.

Socrates' commitment to his work with the Athenians was based in his religious convictions; indeed, he referred to his work as his duty to the god. However, he was a smart and talented man as well as a religious one; although most Athenians condemned his philosophy, some admitted that he was a knowledgeable and skillful teacher.

EFFECTIVE PARAGRAPHING

For the most part, the construction of good paragraphs hinges on all of the aspects of writing discussed thus far. If you think about your idea, brainstorm about specifics, revise sentences into specific, point-by-point details to support your conclusion, and think about the shape of your essay, clearly written paragraphs will seem the most natural way for you to to express your ideas. Some additional guidelines to help you in constructing paragraphs follow.

1. Paragraphs generally work *together;* within an essay, they serve as building blocks on which to rest a conclusion. For example, the second paragraph of the student essay examined earlier begins with a reference to the first paragraph: "In order to understand the Athenians' decision," and then moves into an explanation of the writer's interpretation of Socrates' methods and motivations. If you examine the other paragraphs in this essay (or in any essay), you will

note that most paragraphs make some reference to the paragraphs preceding and following them.

2. Paragraphs also function as units conveying a single point or concept. The indentation at the beginning of a paragraph signals the reader that a premise is going to be presented and supported with specific evidence. The length of the standard paragraph (from 50 to 300 words) is sufficient for the development of a single, illustrated idea. However, this conventional length helps a writer to focus and tighten the idea presented in the paragraph.

3. Finally, paragraphs can be considered essays in miniature. Although paragraphs within an essay are dependent on each other and on the essay's conclusion, each paragraph should contain a single, controlling idea, an illustration of that idea, and finally, a summation of the point or example presented in it. Even beginning and transitional paragraphs, which function to introduce the reader to an idea rather than to explain it, usually follow this general-to-specific format. As you write paragraphs, you will find that keeping this format in mind will help you introduce, explain, and summarize your explanation.

Let's look at one paragraph in isolation in order to better understand its components. The example is the second paragraph of Coretta Scott King's essay, "Why We Still Can't Wait" (*Newsweek*, 1976).

The first reason is that . . . current high unemployment [among African-Americans] is nothing less than a guarantee that America's future will hold deterioration rather than progress. The men and women breadwinners of America are not isolated individuals but a pivot on which the whole health of our community depends. A man with a decent job is the support of his aging parents and a force for stability in his neighborhood and city. Clean and safe streets, decent housing, adequate medical care and even racial peace are all goals that can only come from a base of stable jobs. Without decent jobs, neither the "special programs" of the cautious nor the pious lectures of the uncaring can do anything but add the insult of indifference to the injury of unemployment. Nothing less than the future of America is at stake. Right now a new generation is growing up, all too many in homes where the parent is without work. Tolerating high unemployment in 1980 will be nothing less than a guarantee that we shall walk down dirty streets, past bitter youths and sad-eyed old men, on into the 21st century. High unemployment is nothing less than a vile investment in continuing decay.

If we analyze this paragraph according to the guidelines listed above, we can see several ways in which it adheres to conventions of good paragraphing. First, the phrase *the first reason* relates the paragraph to a previous one so that it functions as a support of a point already raised, but in need of further clarification. Second, the paragraph works to prove one major conclusion: in threatening the prosperity of individuals, unemployment threatens a nation. Every example in the paragraph relates to this point, and the examples follow each other in a logical manner. The paragraph also contains a beginning: "the first reason . . . ," a middle section of specific examples, and an ending that pulls together the conclusion and its supporting statements: "High unemployment is nothing less than a vile investment in continuing decay."

As you construct paragraphs, see if you can revise them so that they reflect these conventions. After you have mastered this format, you can vary your structure a bit. In addition, ask yourself the following questions when you look over a rough draft of your work:

1. Does the paragraph contain one main idea? Does it state this idea with clarity and simplicity?
2. Does all the information in the paragraph support the main idea?
3. Is some information unnecessary?
4. Is some information missing?
5. Do the paragraphs connect with each other? Is there thoughtful and consistent use of logical connectives such as *however, on the other hand, since, because, in addition, in contrast,* and *the first reason?*

EXERCISE

Excerpt three paragraphs from an essay you find particularly interesting (one of the paragraphs should be a beginning to the essay). Using the criteria listed above, analyze each paragraph to discover its main idea, its use of specific details, and ways in which it relates to other paragraphs in the essay. Underline any transitional phrases and any logical connectives you find. Then, write a paragraph that imitates the language and structure of one of the paragraphs you've analyzed. You should not imitate the content of the paragraph; think of another subject and focus on the principles of organization and coherence you've uncovered.

REVISION

As you work on your essay, you will find that much of your activity is centered upon rewriting what you wrote fifteen minutes ago. Revision is absolutely essential to the writing process. Chapter 2 of this book explains that language is conventional, or artificial, and must be learned; the same goes for the essay. You will usually need to construct at least three drafts of your essay if you are to feel comfortable with your final product. Your first draft may only vaguely resemble your intent; it may seem incoherent, full of misspelled and misused words, and too broad in its scope. But that first draft will give you a text you can begin to work with; read it to discover what makes sense and what doesn't, think about the structure you're working toward, and then write a second draft. Your second draft should be read by a friend or classmate, who can help you catch disorganized paragraphs, erroneous examples, and misspelled and misused words. Your third draft may emerge as a reasonably clear one; still, you need to check carefully to see that it is free of typographical and grammatical errors. The choice of how many drafts you write is up to you; keep in mind, however, that you must revise your essay if you are to arrive at a clearly reasoned, clearly written final document.

3. A FINAL NOTE: GRAMMAR AND USAGE

This brief appendix cannot offer specific advice about the parts of speech and about the writing of error-free sentences. Any experiences you may have had with writing essays for your classes has, no doubt, shown the necessity of writing grammatically correct sentences. Instructors may differ widely in their critical evaluations of the manner in which you write arguments, but they all generally want you to observe standard conventions for punctuation, subject-verb agreement, pronoun reference, parallel constructions, among other items in the cornucopia of usage and mechanics. An efficient way for you to learn and apply rules for writing correct sentences is to purchase a handbook for writers and editors. Several options are listed below. Handbooks offer easily accessible, specific definitions of rules governing writing, and often provide exercises to help you apply those rules to your own writing. The books listed below also address the idea that the writing of good sentences, like the writing of good essays, involves thinking about logic and rhetoric as well as about grammar.

SUGGESTED SUPPLEMENTS ON GRAMMAR, USAGE, AND STYLE

Gordon, Karen Elizabeth. *The Well-Tempered Sentence: A Punctuation Handbook for the Innocent, the Eager, and the Doomed.* New York: Ticknor & Fields, 1983.

————. *The Transitive Vampire: A Handbook of Grammar.* New York: Ticknor & Fields, 1984.

Lunsford, Andrea, and Robert Connors. *The St. Martin's Handbook.* New York: St. Martin's, 1989.

Shertzer, Margaret. *The Elements of Grammar.* New York: Macmillan, 1986.

Strunk, William, Jr., and E. B. White. *The Elements of Style,* 3d ed. New York: Macmillan, 1979.

Trimble, John R. *Writing with Style: Conversations on the Art of Writing.* Englewood Cliffs, N.J.: Prentice-Hall, 1975.

Two supplements specific to source-based writing:

Achtert, Walter S., and Joseph Gibaldi. *The MLA Style Manual.* New York: Modern Language Association, 1985.

Gibaldi, Joseph, and Walter S. Achtert. *MLA Handbook for Writers of Research Papers,* 3d ed. New York: Modern Language Association, 1988.

Index